T0248864

Meteorology and Weather Forecasting

Meteorology and Weather Forecasting

Edited by **Dorothy Rambola**

New York

Published by Syrawood Publishing House,
750 Third Avenue, 9th Floor,
New York, NY 10017, USA
www.syrawoodpublishinghouse.com

Meteorology and Weather Forecasting
Edited by Dorothy Rambola

International Standard Book Number: 978-1-68286-002-1 (Hardback)

Printed in the United States of America.

Contents

Preface

Earth's atmosphere is a complex field of study as it involves variables like temperature, air, etc. The technological advances of the twentieth century have contributed majorly towards the progress of meteorology. This comprehensive book examines in detail the current developments, concepts, models, measurement tools, etc. in this field. It will prove beneficial for students, researchers, meteorologists, professionals and researchers alike.

All of the data presented henceforth, was collaborated in the wake of recent advancements in the field. The aim of this book is to present the diversified developments from across the globe in a comprehensible manner. The opinions expressed in each chapter belong solely to the contributing authors. Their interpretations of the topics are the integral part of this book, which I have carefully compiled for a better understanding of the readers.

At the end, I would like to thank all those who dedicated their time and efforts for the successful completion of this book. I also wish to convey my gratitude towards my friends and family who supported me at every step.

Editor

On daily interpolation of precipitation backed with secondary information

S. Krähenmann and B. Ahrens

Institute for Atmospheric and Environmental Sciences, Goethe-University Frankfurt,
Frankfurt am Main, Germany

Abstract. This paper investigates the potential impact of secondary information on rainfall mapping applying Ordinary Kriging. Secondary information tested is a natural area indicator, which is a combination of topographic features and weather conditions. Cross validation shows that secondary information only marginally improves the final mapping, indicating that a one-day accumulation time is possibly too short.

1 Introduction

Rainfall varies in both space and time. This variability increases with shorter time scales. Hence it is more difficult to interpolate with a limited number of observations on daily than on monthly or annual time scales (Haylock et al., 2008; Yatagai et al., 2009).

Many previous approaches have been made to map precipitation from gauge-observations. Some ignored the spatial covariance structure and knowledge of precipitation processes like orographic effects as Thiessen polygons (TP) and inverse distance weighting (IDW) methods do; several considered spatial covariance structure of precipitation as Kriging does (Goovaerts, 2000; Beck and Ahrens, 2006); Goovaerts pointed out that geo-statistical methods, such as Ordinary Kriging (OK) outperform traditional techniques.

In case an external variable is highly correlated to the studied variable, this correlation can be used to improve the spatial interpolation of the variable of interest. A straightforward method to introduce secondary data is by regression of rainfall versus elevation (Daly et al., 1994; Guan and Wilson, 2005). However, rainfall at a particular grid-node is derived from elevation at this point only, not taking into account surrounding point measurements. In Ordinary Co-Kriging (OCK), spatial correlations between the variable of interest and the external variable are used to modify the kriging equation system (Goovaerts, 2000). This method is highly complex as the covariances of all variables have to be estimated together.

However, while precipitation amount is known to increase with terrain height over larger accumulation times, it generally is weakly correlated to terrain height for short (e.g. daily) accumulation times (Daly, 2002; Goovaerts, 2000). Thus instead of including the external information directly into the interpolation algorithm, this study presents a way of considering external information in a modified way regarding weather conditions and greater topographical features.

Subsequently daily precipitation data from Germany are interpolated using two techniques: (1) methods that use only daily rainfall data recorded at 759 stations in Germany (TP, IDW, OK); (2) algorithms that combine rainfall data with secondary information (Kriging of Observational Ratios (KOR), and OcK with natural area indicator (NAI) as secondary information). The prediction performance of the different algorithms is compared using cross validation (e.g., Wackernagel, 2003). Stochastic interpolation, which is related to Kriging, but not minimizing the Kriging-variance, is used for evaluating the quality of estimation uncertainty provided by techniques implementing ancillary information.

2 Data

Germany consists of a complex topography, ranging from the flat maritime area to the north close to sea level, the hilly low mountain ranges in the middle part with terrain heights of 500 to 1000 m and the alpine area to the south, with an area of approximately 357 000 km². To the south west the Black Forest and Swabian Mountains are most prominent with up to 1500 m, in the south-eastern part close to the border to Austria the elevation reaches almost 3000 m.

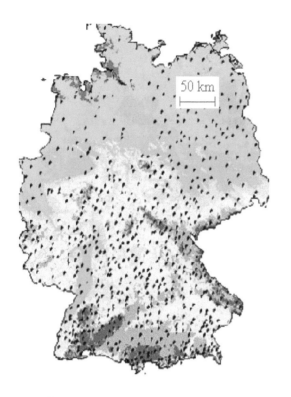

Figure 1. Orographical map of Germany with 759 daily operationally measuring rain-gauges indicated by black dots. Low areas green, and high areas brown.

2.1 Rain-gauge network

For this study 759 operationally measuring rain-gauges were available, operated by the German Weather Service (DWD), indicated by black dots in Fig. 1. They provide a network of continuous measurements with a fairly homogenous nationwide coverage, and an average distance between neighbouring stations of about 21 km. However, the density of the stations is reduced in higher elevations. Due to the height dependence of rainfall distribution, such biases in the station distribution can lead to systematic errors in the interpolation procedure.

2.2 Secondary information

Terrain elevation is known to be highly correlated with climate variables at least over longer accumulation times (Goovaerts, 2000; Daly et al., 2002). Here, a digital elevation model (DEM) is used with a spatial resolution of 1 km^2 as secondary information (available from US Geological Survey, EROS Data Center, Sioux Falls, SD). The orography is shown in Fig. 1. From the DEM further information is deduced, such as steepness of terrain, aspect ratio and prominence of terrain.

2.3 Weather types

The propagation of rain-patterns is highly influenced by the interaction of orography and the wind-field. For example, mountain ridges are able to partially block moisture bearing air, forcing it to ascend. This leads to pronounced precipitation on windward sides, and less rain or even no rain on leeward sides of terrain. Especially in case of low wind speeds blocking is more likely, as the ability of moist air to rise above mountain ridges and propagation of rainfall towards lee-ward sides is limited.

This study distinguishes between nine weather types, characterized by wind direction and wind speed. The four main wind directions (NW, SW, SE, NE) are further split into strong wind and weak wind by a threshold of 10 m/s, the ninth weather type is characterized by no prevailing wind direction.

The horizontal wind direction for each grid point is derived from the wind component data of the 700 hPa level (Bissolli and Dittmann, 2001), distinguishing the four main wind directions. Daily average wind speed, with a spatial resolution of 2.5° × 2.5°, averaged over Germany, at the 850 hPa level, is noted by http://www.esrl.noaa.gov/psd/data/gridded/tables/land.html.

3 Methods

This section briefly introduces the different interpolation methods used in this study.

3.1 Univariate methods

3.1.1 Thiessen Polygone (TP) Method

This is a simple interpolation method assigning to each grid cell the value of the closest observation and is also called nearest neighbour interpolation (Goovaerts, 2000). An example is given in Fig. 2a, representing daily rainfall on 1.1.2007.

3.1.2 Inverse Distance Weighting (IDW)

To prevent unrealistic artefacts at polygon borders, rainfall can be estimated as a linear combination of surrounding rain-gauge observations, with the weights being inversely proportional to the distance between observations to the power p. The idea of the weighting system is to put more emphasis on the observations closest to the grid cell to be estimated (Wackernagel, 2003). In this study interpolation is done from 6 surrounding observations and a distance weighting power of $p = 1.6$ yielding the least RMSE error using cross validation with all stations in Germany for all wet days in 2007, and was held constant over the whole year.

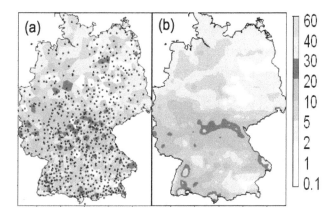

Figure 2. Rain maps for Germany, 1.1.2007 (units in mm). Used interpolation methods are (**a**) Thiessen Polygones (TP) and (**b**) Ordinary Kriging (OK).

3.2 Geostatistical methods without ancillary variables

3.2.1 Ordinary Kriging (OK)

OK is a generalized least-square regression technique that allows to account for spatial dependence between observations. Like inverse distance weighting OK estimates the unknown rain amount at grid cells as a linear combination of neighbouring observations. The weights are obtained by minimization of the estimation variance, while ensuring the unbiasedness of the estimator. Instead of Euclidean distance, OK uses a semivariogram as a measure of distance in the observed rain-field. The semivariogram reflects the intuitive feeling that measurements of two rain-gauges close to each other are more alike than those further apart. Like IDW and TP the OK (without nugget effect) reproduces the observations at the station locations.

A spherical climatological variogram model with range of 160 km is chosen for interpolating daily rainfall. Input data firstly are normal score transformed, afterwards all ranges are averaged for days with at least 0.1 mm precipitation per day German wide. This has two reasons, on the one hand this yields the least RMSE in cross validation, on the other hand the stochastic simulation procedure becomes more stable. An example is given in Fig. 2b, representing daily rainfall on 1.1.2007.

3.3 Geostatistical methods including secondary information

The incorporation of secondary information potentially improves the estimation of the true rain-field. A straightforward approach is to predict rainfall as a function of the collocated elevation, where elevation data are available at all estimation grid-nodes. The foremost disadvantage of this approach is that the rainfall-amount at a particular grid-node is derived from the elevation only, regardless of the measurements at surrounding rain-gauges. This approach presumes that the residual values are spatially uncorrelated. A more promising approach is to combine a geostatistical method, which is able to account for spatial correlation of rainfall, with ancillary information. In this study two types are tested: (1) Ordinary Co-Kriging (OCK), and (2) Kriging of Observational Ratios (KOR).

3.3.1 Ordinary Co-Kriging (OCK)

The OCK is a multivariate extension of OK. In OCK, spatial correlations between the variable of interest and the secondary variable are used to modify the Kriging system (Goovaerts, 2000). Disadvantageous is its screening of further away elevation data. Furthermore the Co-Kriging system can be unstable in case of inhomogeneous orographical structures, as it is the case in Germany. The main reason for this instability is the much higher correlation between close elevation data, than the correlation between distant rainfall data (Goovaerts, 2000; Wackernagel, 2003). To avoid instability in the subsequent modelling of direct and cross-semivariograms, elevation is computed from the 759 rain-gauge stations only, not the entire DEM. Details of elevation map do not appear in the rainfall-map as elevation is only taken for improving the estimate of spatial variability.

3.3.2 Kriging of Observational Ratios (KOR): natural area indicators as secondary information

This is an alternative method to OCK, which decouples the regression and the interpolation part. Here, OK is performed on ratios obtained dividing daily observations by a natural area indicator. Ratios have the advantage of avoiding negative values as well as allowing for non-linear relations between primary and secondary variable.

Primarily an indicator value specific to the prevailing weather condition is calculated for each grid-node, combining orographical parameters deduced from a DEM (elevation, slope, prominence of terrain, and wind facing direction), which is explained in the following. The weather specific indicator value links the orography to the observed rainfall (averaging rainfall for similar conditions). Thereafter the natural area indicator is calculated, such as to maximize the correlation between a single day observation and his respective indicator value specific to the prevailing weather conditions.

Orographical parameters

To only account for main orographical features, elevation data (h) are smoothed using a Gaussian filter with radius 9 km (Fig. 3a), inspired by a study of Smith et al. (2003) which found that the dominant spatial-scale of lifting and rainfall is in the Alps about 10 km.

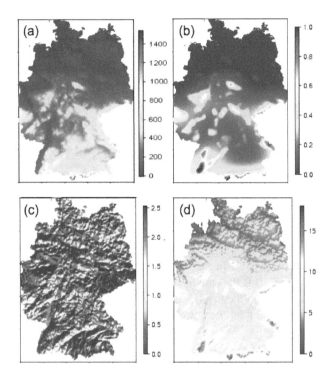

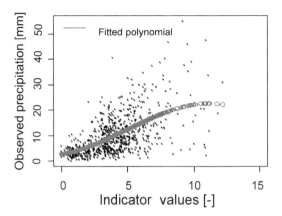

Figure 4. Precipitation versus natural area indicator for 1.1.2007. A 3rd order fit, which maximizes the correlation between trend-value and observation is shown (correlation coefficient = 0.577).

Figure 3. Illustration of terrain features deduced from the DEM: **(a)** smoothed DEM using Gaussian-filter with radius 5 km [m], **(b)** prominence of terrain, **(c)** relative facing-value with wind from NW, wind speed >10 m/s, **(d)** indicator values [–].

Slope is calculated as the maximum change in elevation over the distance between the cell and its eight neighbours.

The aspect depicts the down-slope direction of the maximum rate of change in elevation from each node to its neighbours, and is useful for distinguishing windward from leeward sides as a function of wind direction, and wind speed. For each single day a facing-value is deduced from the aspect ratio given the mean wind direction, and the wind speed. The relative down-weighting of the area on the leeward side is smaller in case wind-speed exceeds 10 m/s (Fig. 3c). The weights are calculated, such as to maximize the correlation between observed rainfall and facing-value.

The so called prominence (prom) of terrain specifies the effectiveness of terrain to alter the precipitation-field. According the method suggested by Daly et al. (2002) for each grid node the maximum elevation difference to grid nodes within radius 100 km is determined Grid nodes with elevation differences greater than 500 m are assigned 1, smaller than 100 m 0, or otherwise a value between 0 and 1 (Fig. 3b).

Calculation of the indicator

Nine climatological indicator values are determined, one for each weather type. Observations are averaged for days of the same weather type, yielding nine climatological observations (obs.clim) for the k rain-gauges:

$$\text{obs.clim}_{i,j,k} = \frac{1}{N} \sum_{n=1}^{N} \text{obs}_{i,j,k} \qquad (1)$$

where N is the number of days of the same weather type, i the wind direction (NW, SW; SE; NE; no wind), and j the wind speed (above or below 10 m/s).

The indicator values are calculated combining previously defined orographical parameters:

$$\text{Ind}_{i,j} = \log(\text{slope } h)\, \text{prom} + \log(\text{facing}_{i,j}\, h) \qquad (2)$$

such as to maximize the correlation to the respective climatological observations at the k locations (Fig. 3d):

$$\text{cor}(\text{Ind}_{i,j,k}, \text{mean}(\text{obs.clim}_{i,j,k})) = \max \qquad (3)$$

Finally a natural area indicator (NAI) is calculated, which is adjusted to daily rain-gauge observations. A non-linear regression line (3rd order polynomial) defines daily rain-gauge observations as a function of indicator values of the respective weather condition (Fig. 4):

$$\text{obs}_k \sim a + b\,\text{Ind}_k + c\,\text{Ind}_k^2 + d\,\text{Ind}_k^3 \qquad (4)$$

yielding parameters a, b, c, and d. These parameters are applied on indicators to calculate the NAIs:

$$\text{NAI} = a + b\,\text{Ind} + c\,\text{Ind}^2 + d\,\text{Ind}^3 \qquad (5)$$

This allows for non-linear relations between daily rainfall and indicator value. The rain-gauge observations are divided by the NAI values at the k locations:

$$\text{ratio}_k = \frac{\text{obs}_k}{\text{NAI}_k} \qquad (6)$$

yielding k ratios, which are normalized before interpolation is done.

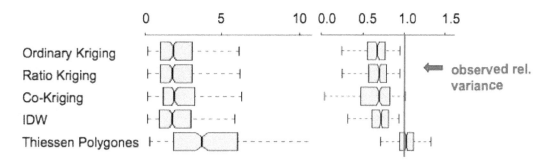

Figure 5. Comparison of different interpolation methods using cross validation. The box plots indicate RMSE (left), and relative variances (right) found for all days in 2007.

3.4 Modelling uncertainty

The probabilistic way to model the uncertainty of a variable at any grid-node consists of viewing the unknown value as the realization of a random variable, and deriving its conditional cumulative distribution function (ccdf). The ccdf fully models the uncertainty at a grid-node, since it gives the probability that the unknown variable is now greater than any given threshold. Under the multiGaussian model, the ccdf at any location is Gaussian and completely characterized by its mean and variance, which corresponds to the Kriging estimate and variance. The approach requires a prior normal score transform of the input data to ensure that at least the univariate distribution is normal (Deutsch and Journel, 1998). The normal score ccdf then is back-transformed to yield the ccdf of the original variable.

Stochastic simulation (SGI, Deutsch and Journel, 1998; Ahrens and Beck, 2008) has been used to assess local uncertainty from the local distribution of simulated values; that is the ccdf at u is approximated by:

$$F(u;z|(n)) \approx \frac{1}{L} \sum_{l=1}^{L} i^{(l)}(u;z) \qquad (7)$$

where $i^{(l)}(u;z) = 1$ if $z^{(l)}(u) \leq z$ and 0 otherwise, with z the simulated value, n the neighbouring data, and L the number of simulations. In theory, as the number of realizations tends to infinity, the local distribution of simulated values should match that provided by Kriging within a similar framework (Goovaerts, 2001).

4 Results

The performance of the different interpolation methods is assessed and evaluated using cross validation. The idea behind cross validation is to remove each rain-gauge observation once in turn from the input dataset and to re-estimate the rainfall-amount from the remaining dataset using an interpolation method. The evaluation-criteria are the root mean square of the prediction error (RMSE, perfect value 0), and

the variance of the interpolated rain-field relative to the variance of the input data (perfect value equals 1).

The relative variance reflects the ability of the different interpolation algorithms to maintain the spatial variability within the rain-field. The relative variance is calculated as the ratio of the estimated variability and the true variability given the left-out rain-gauges.

The evaluation is summarized in Fig. 5. The box plots for RMSE indicate the range of interpolation errors for all wet days (at least 0.1 mm station mean) in 2007. The IDW and the three Kriging methods (OK, OCK, KOR) perform equally well, while Thiessen Polygones yields a clearly larger RMSE value.

In terms of relative variance Thiessen Polygones yield the best score. Hence, as the method can not distinguish between close by rain gauges (more alike) and further away rain-gauges (less similar), cross-validation yields a large RMSE, but highlights its strength in maintaining the spatial variability. IDW performs on average slightly better than the three Kriging methods (Fig. 5). However, IDW does not provide a direct uncertainty measure as geostatistical algorithms do (Kriging, SGI).

Of particular interest is the result for OCK in terms of relative variance, the large spread of the box plot suggests that OCK retains the spatial variability in some cases better but many of the cases worse than OK. Overall KOR and OCK do not outperform OK, neither in terms of RMSE nor in terms of relative variance. This indicates that the gained value, if there is any, does not appear in this evaluation exercise and might be a consequence of the generally low correlation coefficient (e.g. 1.1.2007, Fig. 4).

Different methods yield models of uncertainty that can greatly differ, and a legitimate question is weather the choice of a technique (e.g. OK, OCK, KOR) can be supported by the data. Cross-validation can be used to build uncertainty models which are then compared with observations that have been temporarily removed one at a time.

At any test location u, knowledge of the ccdf $F(u;z|(n))$ allows the computation of a series of symmetric p-probability intervals (PI) bounded by the $(1-p)/2$ and $(1+p)/2$ quantiles

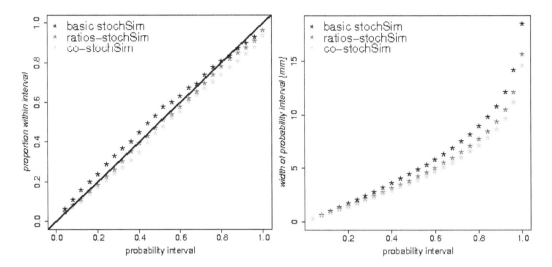

Figure 6. Accuracy plot indicating the ability of the different interpolators (OK, OCK, KOR) to estimate the true width of uncertainty measure (left), and width of uncertainty measure for PIs (right) for operational rain-gauges in Germany on 1.1.2007.

of that ccdf. For example, the 0.5-PI is bounded by the lower and upper quantiles $[F^{-1}(u;0.25|(n)), F^{-1}(u;0.75|(n))]$. A correct modelling of local uncertainty would implicate that there is a 0.5 probability that the actual z-value at u falls into that interval or, equivalently, that over the study area, 50% of the 0.5-PI include the true value. If a set of z-measurements and independently derived ccdfs are available at N locations, $u_j, \{[z(u_j), F(u_j, z|(n))]\}$, $j = 1,...N$, the fraction of true values falling into the symmetric p-PI can be computed as:

$$\bar{\xi}(p) = \frac{1}{N} \sum_{j=1}^{N} \xi(u_j; p) \; \forall \; p \in [0,1] \qquad (8)$$

with $\xi(u_j; p) =$

$$\begin{cases} 1 & \text{if } F^{-1}(u_j;(1-p)/2) < z(u_j) \leq F^{-1}(u_j;(1+p)/2) \\ 0 & \text{otherwise} \end{cases} \qquad (9)$$

Here, the ccdfs are inferred through 100 SGI (OK-, OCK-, KOR- equation).

The scattergram of the estimated versus expected fractions is called "accuracy plot". Figure 6 (left) shows the daily rainfall accuracy plot computed for both OK, OCK, and KOR ccdfs using cross-validation of 759 observations at all wet days in 2007. The accuracy plot shows that OK overestimates the uncertainty for PI smaller than 0.6 (black stars above black line), e.g. the 0.5 PI derived from the OK contains 60% of the true values, while KOR, and OCK perform better. For PIs greater than 0.6 OK performs better. Here, COK contains too few true values.

5 Conclusions

Secondary information has to be considered in terms of accumulation time and complexity of terrain. Combining OK with a NAI only slightly improves the evaluation results independent of the applied combination approach. Introduction of secondary information with observational ratios yields a small improvement in terms of uncertainty measure, as could be shown with the accuracy plot (Fig. 6, left).

As is shown, the number of true values within interval is closer to the expected number (given by the PI) when incorporating NAI values. Otherwise the width of the probability interval seems too wide indicating overestimation of the true uncertainty. This is not true for probability intervals greater than 0.6. Overall the gain is minimal. The foremost reason for this is the too short accumulation time, as the spatial distribution of precipitation tends to be better defined by topographical/external parameters for longer accumulation times (Goovaerts, 2000). New techniques introducing secondary information should be investigated, e.g. by means of a stratified variogram which is separately inferred from rain-gauges within similar topographical features. Another way might be to determine parameters such as NAI on a physically meaningful basis, or adding further parameters such as humidity or stability (Haiden et al., 2008), instead of determining them statistically. The main challenge is the fact that any simple, static, either topographical or weather-type based index can capture only a small part of the complex processes involved.

Acknowledgements. Precipitation data were obtained from the German Weather Service (DWD). The authors also acknowledge funding from the Hessian initiative for development of scientific and economic excellence (LOEWE) at the Biodiversity and Climate Research Centre (BiK-F), Frankfurt/Main.

Edited by: M. Dolinar
Reviewed by: T. Haiden and another anonymous referee

References

Ahmed, S. and de Marsily, G.: Comparison of geostatistical methods for estimating transmissivity using data on transmissivity and specific capacity, Water Resour. Res., 23, 1717–1737, 1987.

Ahrens, B.: Distance in spatial interpolation of daily rain gauge data, Hydrol. Earth Syst. Sci., 10, 197–208, 2006, http://www.hydrol-earth-syst-sci.net/10/197/2006/.

Ahrens, B. and Beck, A.: On upscaling of rain-gauge data for evaluating numerical weather forecasts, Meteorol. Atmos. Phys., 99(3–4), 155–167, 2008.

Bissolli, P. and Dittmann, E.: The objective weather type classification of the German Weather Service and its possibilities of application to environmental and meteorological investigations, Meteorol. Z., 10(4), 253–260, 2001.

Daly, C., Neilson, R. P., and Phillips, L. D.: A statistical-topographic model for mapping climatological precipitation over mountainous terrain, J. Appl. Meteorol., 33, 140–158, 1994.

Daly, C., Gibson, W. P., Taylor, G. H., Johnson, G. L., and Pasteris, P.: A knowledge-based approach to the statistical mapping of climate, Clim. Res., 22, 99–113, 2002.

Daly, C.: Variable influence of terrain on precipitation patterns: Delineation and Use of Effective Terrain Height in PRISM, Oregon State University, Corvallis, available at: http://www.prism.oregonstate.edu/pub/prism/docs/effectiveterrain-daly.pdf, 2002.

Deutsch, C. V. and Journel, A. G. (Eds.): GSLIB: Geostatistical Software Library and User's Guide, Oxford Univ. Press, Oxford, UK, 369 pp., 1998.

Goovaerts, P.: Geostatistical approaches for incorporation elevation into the spatial interpolation of rainfall, J. Hydrol., 228, 113–129, 2000.

Goovaerts, P.: Geostatistical modelling of uncertainty in soil science, Geoderma, 103, 3–26, 2001.

Guan, H. and Wilson, J. L.: Geostatistical Mapping of Mountain Precipitation Incorporating Autosearched Effects of Terrain and Climatic Characteristics, J. Hydrometeorol., 6(6), 1018–1031, 2005.

Haiden, T. and Pistotnik, G.: Intensity-dependent parameterization of elevation effects in precipitation analysis, Adv. Geosci., 20, 33–38, 2009, http://www.adv-geosci.net/20/33/2009/.

Haylock, M. R., Hofstra, N., Klein Tank, A. M. G., Klok, E. J., Jones, P. D., and New, M.: A European daily high-resolution gridded dataset of surface temperature and precipitation, J. Geophys. Res., 113, D20119, doi:10.1029/2008JD010201, 2008.

Pebesma, E. J.: Multivariable geostatistics in S: the gstat package, Comput. Geosci., 30, 683–691, 2004.

Prudhomme, C. and Reed, D. W.: Mapping of extreme rainfall in a mountainous region using geostatistical techniques: A case study in Scotland, Int. J. Climatol., 19, 1337–1356, 1998.

Smith, R. B., Quingfang, J., Fearun, M. G., Tabary, P., Dorninger, M., Doyle, J. D., and Benoit, R.: Orographic precipitation and air mass transformation: An Alpine example, Q. J. R. Meteorol. Soc., 129, 433–454, 2003.

Wackernagel, H. (Eds.): Multivariate Geostatistis, 3rd Edn., Springer, Berlin, 2003.

Yatagai, A., Arakawa, O., Kamaguchi, K., Kawamato, H., Nodzu, M. I., and Hamada, A.: A 44-year daily gridded precipitation dataset for Asia based on a dense network of rain gauges, SOLA: Scientifoc Online Letters of the Atmosphere, 5, 137–140, doi:10.2151/sola.2009-035, 2009.

2

Development of a longterm dataset of solid/liquid precipitation

B. Chimani, R. Böhm, C. Matulla, and M. Ganekind

Central Institute for Meteorology and Geodynamics, Vienna, Austria

Abstract. Solid precipitation (mainly snow, but snow and ice pellets or hail as well), is an important parameter for climate studies. But as this parameter usually is not available operationally before the second part of the 20th century and nowadays is not reported by automatic stations, information usable for long term climate studies is rare. Therefore a proxy for the fraction of solid precipitation based on a nonlinear relationship between the percentage of solid precipitation and monthly mean temperature was developed for the Greater Alpine Region of Europe and applied to the existing longterm high resolution temperature and precipitation grids (5 arcmin). In this paper the method is introduced and some examples of the resulting datasets available at monthly resolution for 1800–2003 are given.

1 Introduction

The ratio of solid to total precipitation has been part of the routine meteorological observing and digitizing programmes of weather services for several decades of the 20th century. Back into the 19th century and further back into the early instrumental period (e.g. Böhm et al., 2010) such data are not available at daily resolution and least of all in homogenized form (e.g. Moberg et al., 2006). In recent decades automation of the networks makes the splitting of precipitation into its solid and liquid parts vanishing more and more from the observing practice due to typical automated networks without observers. Therefore an increasing lack of such parameters which are either not or expensive to automate is going to cause deficiencies for research and practical applications in the future. Nevertheless solid precipitation is an important climate parameter in terms of practical applications and a parameter sensitive to the ongoing and anticipated warming.

In this paper we want to introduce a method to produce "instrumental based proxies" for the solid/liquid precipitation fractions and for totals of solid/liquid precipitation derived from monthly mean temperatures and monthly precipitation totals. The derived method is applied on the existing high-resolution longterm grids of these two basic climate parameters (Efthymiadis et al., 2006; Chimani et al., 2010) in the "Greater Alpine Region" (henceforth GAR), which

reaches from 4°–19° E and 43°–49° N (Auer et al., 2005, 2007). The resulting new datasets provide monthly, seasonal and annual solid/liquid precipitation totals back to 1800 and the monthly ratios of solid/liquid precipitation back to 1780 – at a spatial resolution of 5 arcmin which may be the maximum possible resolution in terms of the given network density. This resolution is adequate for a considerable number of scientific and practical applications in the topographically complex terrain of the study region.

As a positive side effect the method of "instrumental based proxies" – if applied on well homogenizable monthly input data – produces "indirectly homogenized" time series of climate parameters elsewise hardly homogenizable directly.

In Sect. 2 the method for deriving the percentage and amount of solid precipitation is introduced, Sect. 3 shows some results achieved with this new dataset. The conclusions are summarized in Sect. 4.

2 Method

2.1 Derivation of the relationship between monthly mean temperature and the percentage of solid precipitation

Daily measurements of temperature, precipitation amount and precipitation type of more than 170 stations in Austria for 1950 to 2008 were used to analyze and describe a relationship between the mean monthly temperature and the monthly percentage of solid precipitation. This was done by determining a hyperbolic tangent. A hyperbolic tangent is a suitable type of function when a parameter with an upper and

Table 1. Root mean square error (RMSE) of the calculated to the measured percentages of solid precipitation for each month over the period 1950–2008.

	RMSE
January	24.99
February	21.84
March	20.83
April	14.97
May	9.09
June	5.93
July	5.29
August	4.93
September	5.85
October	12.23
November	21.13
December	24.26

a lower limit has to be regressed versus an unlimited one. In our case the percentage of solid precipitation is variable between 100% (at lower temperatures) and 0% (at higher temperatures). A similar approach has been successfully applied already on snowcover duration vs. temperature (Hantel et al., 2000; Wielke et al., 2004). The function of the form

$$p = 50 \tanh(2\, a_1\, (T - a_2)) + 50 \qquad (1)$$

with p as solid precipitation fraction in %, T as temperature in °C, and a_1 and a_2 being the factors describing the maximum slope and the offset of the function from 0 °C at the inflection point. The regressions were determined for each single month. A further splitting of the data for example into different climate regions or height layers did not improve the relationship significantly but caused problems due to too few data in some of the resulting classes. The function was determined by minimizing weighted differences of monthly mean temperature and the percentage of solid precipitation (Eqs. 2 and 3) between the function and the measurements.

$$w\, \frac{\mathrm{MSE}_{RR}}{\Sigma(\mathrm{MSE})}\, \mathrm{MSE}_T + (1-w)\, \frac{\mathrm{MSE}_T}{\Sigma(\mathrm{MSE})}\, \mathrm{MSE}_{RR} \to \mathrm{Min} \qquad (2)$$

$$w = \frac{n_{>10}}{n_{\mathrm{all}}} \qquad (3)$$

w is the weighting factor determined by the number (n) of the measurements having more than 10% of solid precipitation and the number of all measurements, MSE the mean squared error for temperature ($_T$) and percentage of solid precipitation ($_{RR}$) and $\Sigma(\mathrm{MSE})$ the sum of both.

The weighting was necessary due to the different distributions of the percentage of solid precipitation in the different months, especially as in the summer months only low values of solid precipitation occur. The RMSE (root mean square error) of the function for all months can be seen in Table 1. Its maximum lies in January with about 25% and the minimum in August with 5%. This is due to the greater variability in the data during winter. It can be noticed by Fig. 1 (left

side) that the functions determined for winter months have a steeper slope than the ones for summer times. The right hand side of Fig. 1 shows the data and the fitted tanh-function for January.

2.2 Generating the longterm dataset of solid precipitation

In order to gain high resolution monthly fields of the percentage of solid precipitation for the last 200 years for GAR the monthly hyperbolic tangents of Sect. 2.1 were applied to the recently produced long term temperature dataset in the region. This dataset covers a period from 1780 to 2008 at a resolution of 5 arcmin. It was created by blending highly resolved monthly mean temperature grids 1961–1990 of the region (Hiebl et al., 2009) with longterm lower resolution anomaly fields in two different height layers to account for the decoupling of atmospheric layers e.g. due to inversions. To provide a comparable spatial resolution of the data over the whole period missing parts of the station-mode series were reconstructed by an EOF-method. For details see Chimani et al. (2010).

Moreover a regional high resolution longterm absolute monthly precipitation dataset was used to generate information not only about the percentage of solid precipitation over the last 200 years but also for the solid/liquid totals in mm. For details of the underlying precipitation dataset see Efthymiadis et al. (2006). It covers the period from 1800 to 2003. This dataset covers the same region as the long term temperature dataset and had only to be rescaled from 10′ to the aspired 5′ grid resolution. This rescaling was done by a straightforward approach without introducing additional information. As no precipitation dataset, developed on a 5-min grid, was available for the whole GAR the additional uncertainties had to be accepted.

3 Results

Figure 2 shows an example of the resulting fields of the percentage and the amount of the solid precipitation for December 1929. In the image of the percentage of solid precipitation (Fig. 2a) the high percentage on the Alpine ridge, due to locally low temperatures, can be seen. The Apennine, the Dinaric Alps and the Jura can also be identified easily. Moreover the continental influence in the eastern part of the GAR compared to the western part can be noticed by the smaller percentage of solid precipitation in the west and especially near the Mediterranean. The structure of the amount of solid precipitation (in mm water equivalent) (Fig. 2b) shows a similar picture with a higher amount of solid precipitation in the mountainous regions but especially north of the Alps the areas of higher solid precipitation amount extends farther north than the percentage alone would explain. Here the influence of the precipitation distribution is visible.

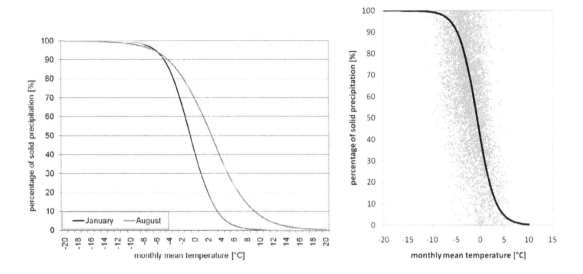

Figure 1. Left: Hyperbolic tangent function describing the relationship between the monthly mean temperature [°C] (abscissa) and the percentage of solid precipitation [%] (ordinate) for January (black line) and August (dark grey line). Thin light grey lines show the functions for all the other months. Right: Example of the function fitting for January. The black line shows the fitted function and the grey dots the measurements.

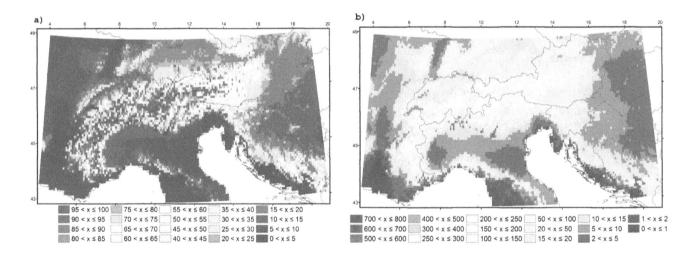

Figure 2. Percentage [%] (**a**) and amount [mm water equivalent] (**b**) of solid precipitation in the GAR for December 1929. Red/blue colours representing high/low values.

Figure 3 displays time series of 11 year running means of the totals and of the percentages of solid winter precipitation of some grid points in the GAR. Nearly no longterm trend in the percentage of solid precipitation can be seen for the inneralpine city of Innsbruck whereas the totals of solid precipitation show marked decadal variations with periods of small amounts of solid precipitation near 1855, 1935 and 1970. The lowest values can be found during the 1855 minimum. The series of the lower and extra-alpine city of Vienna however show opposite features: Higher decadal scale variablility and also longterm decreasing trend of the per-

centage of solid precipitation and smoother time series of the solid precipitation totals. Another interesting structure shown by these plots is the higher percentage of solid precipitation for south-inner-alpine Bozen, compared to the relatively low solid precipitation fractions in the milder northern-extra-alpine Zürich, a site with significantly more total precipitation and therefore comparable sums of solid precipitation.

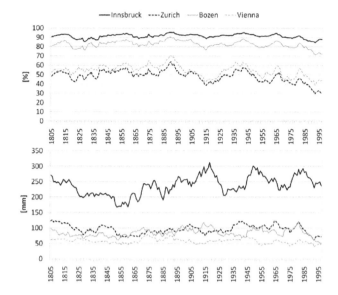

Figure 3. Time series of the 11 year running mean for the percentage [%] (upper plot) and amount [mm water equivalent] (lower plot) of solid precipitation for winter seasons (DJF). Displayed are the time series of 4 gridpoints located near the cities Innsbruck (black solid line), Zurich (black broken line), Bozen (grey solid line) and Vienna (grey broken line).

4 Conclusions

A method was developed to overcome the lack of information about solid precipitation in former centuries and due to the recent automation of measuring networks. The method is based on high resolution monthly temperature and precipitation data and a monthly relationship between mean temperature and the percentage of solid precipitation. We introduced the term "instrumental proxies" for such a type of climate parameters which for longer timescales are not existing directly but maybe indirectly derived from other main climate parameters available in homogenized form for the entire instrumental period. In our case we achieve monthly information for both, the percentage and the amount of solid precipitation for the Greater Alpine Region for the period 1800 to 2003. The spatial resolution is 5 arcmin.

As shown at a few examples here there is a wide field of potential scientific and practical applications using either a map-based approach to study the spatial distribution of solid precipitation or a time series approach to study longterm trends and decadal to annual scale variability for single sites as well as for subregions selected according to specific objectives.

In our opinion the relationship between solid precipitation ratios and temperature should be robust to spatial variations. Nevertheless it might be advisable to define an adequate relationship between monthly mean temperature and the percentage of solid precipitation for different regions as regional specifics like inversion layers could influence the results. As it is necessary to have measurements of the solid precipitation to define the relationship, this might lead to some problems.

Acknowledgements. The dataset was developed in the context of the internal project GAR-SCHNEE-200 of the Central Institute for Meteorology and Geodynamics, Vienna, funded by the Austrian Ministry for Science and Research.

Edited by: M. Dolinar
Reviewed by: two anonymous referees

References

Auer, I., Böhm, R., Jurković, A., Orlik, A., Potzmann, R., Schöner, W., Ungersböck, M., Brunetti, M., Nanni, T., Maugeri, M., Briffa, K., Jones, P., Efthymiadis, D., Mestre, O., Moisselin, J.-M., Begert, M., Brazdil, R., Bochnicek, O., Cegnar, T., Gajić-Čapka, M., Zaninović, K., Majstorović, Ž., Szentimrey, T., and Mercalli, L.: A new instrumental precipitation dataset for the Greater Alpine Region for the period 1800–2002, Int. J. Climatol., 25, 139–166, 2005.

Auer, I., Böhm, R., Jurkovic, A., Lipa, W., Orlik, A., Potzmann, R., Schöner, W., Ungersböck, M., Matulla, C., Briffa, K., Jones, P., Efthymiadis, D., Brunetti, M., Nanni, T., Maugeri, M., Mercalli, L., Mestre, O., Moisselin, J.-M., Begert, M., Müller-Westermeier, G., Kveton, V., Bochnicek, O., Stastny, P., Lapin, M., Szalai, S., Szentimrey, T., Cegnar, T., Dolinar, M., Gajic-Capka, M., Zaninovic, K., Majstorovic, Z., and Nieplova, E.: HISTALP—Historical instrumental climatological surface time series of the Greater Alpine Region 1760–2003, Int. J. Climatol., 27, 17–46, 2007.

Böhm, R., Jones, P. D., Hiebl, J., Frank, D., Brunetti, M., and Maugeri, M.: The early instrumental warm-bias: a solution for long central European temperature series 1760–2007, Clim. Change, 101, 41–67, 2010.

Chimani, B., Matulla, C., Böhm, R., and Hofstätter, M.: A new high resolution absolute Temperature Grid for the Greater Alpine Region back to 1780, Int. J. Climatol., submitted, 2010.

Efthymiadis, D., Jones, P. D., Briffa, K. R., Auer, I., Böhm, R., Schöner, W., Frei, C., and Schmidli, J.: Construction of a 10-min-gridded precipitation data set for the Greater Alpine Region for 1800–2003, J. Geophys. Res., 111, D01005, doi:10.1029/2005JD006120, 2006.

Hantel, M., Ehrendorfer M., and Haslinger A.: Climate sensitivity of snow cover duration in Austria, Int. J. Climatol., 20, 615–640, 2000.

Hiebl, J., Auer, I., Böhm, R., Schöner, W., Maugeri, M., Lentini, G., Spinoni, J., Brunett, M., Nanni, T., Tadić, M. P., Bihari, Z., Dolinar, M., and Müller-Westermeier, G.: A high-resolution 1961–1990 monthly temperature climatology for the greater Alpine region, Meteorol. Z., 18(5), 507–530, 2009.

Moberg, A., Jones, P. D., Lister, D., Walther, A., Brunet, M., Jacobeit, J., Alexander, L. V., Della-Marta, P. M., Luterbacher, J., Yiou, P., Chen, D., Klein Tank, A. M. G., Saladié, O., Sigró, J., Aguilar, E., Alexandersson, H., Almarza, C., Auer, I., Barriendos, M., Begert, M., Bergström, H., Böhm, R., Butler, C. J., Caesar, J., Drebs, A., Founda, D., Gerstengarbe, F.-W., Micela, G., Maugeri, M., Österle, H., Pandzic, K., Petrakis, M., Srnec, L., Tolasz, R., Tuomenvirta, H., Werner, P. C., Linderholm, H., Philipp, A., Wanner, H., and Xoplaki, E.: Indices for daily temperature and precipitation extremes in Europe analyzed for the period 1901–2000, J. Geophys. Res., 111, D22106, doi:10.1029/2006JD007103, 2006.

Wielke, L.-M., Haimberger, L., and Hantel, M.: Snow cover duration in Switzerland compared to Austria, Meteorol. Z., 13, 13–17, 2004.

3

Data-driven exploration of orographic enhancement of precipitation

L. Foresti[1], M. Kanevski[1], and A. Pozdnoukhov[2]

[1]Institute of Geomatics and Analysis of Risk, University of Lausanne, Switzerland
[2]National Centre for Geocomputation, National University of Ireland Maynooth, Ireland

Abstract. This study presents a methodology to analyse orographic enhancement of precipitation using sequences of radar images and a digital elevation model. Image processing techniques are applied to extract precipitation cells from radar imagery. DEM is used to derive the topographic indices potentially relevant to orographic precipitation enhancement at different spatial scales, e.g. terrain convexity and slope exposure to mesoscale flows. Two recently developed machine learning algorithms are then used to analyse the relationship between the repeatability of precipitation patterns and the underlying topography. Spectral clustering is first used to characterize stratification of the precipitation cells according to different mesoscale flows and exposure to the crest of the Alps. At a second step, support vector machine classifiers are applied to build a computational model which discriminates persistent precipitation cells from all the others (not showing a relationship to topography) in the space of topographic conditioning factors. Upwind slopes and hill tops were found to be the topographic features leading to precipitation repeatability and persistence. Maps of orographic enhancement susceptibility can be computed for a given flow, topography and forecasted smooth precipitation fields and used to improve nowcasting models or correct windward and leeward biases in numerical weather prediction models.

1 Introduction

The orographic precipitation enhancement is a complex atmospheric phenomenon which is the subject of many numerical (Rotunno and Houze, 2007) and observational studies (Gray and Seed, 2000; Panziera and Germann, 2010). High-resolution numerical weather prediction (NWP) models are computationally demanding to provide fast forecasts with appropriate data assimilation systems. Expert-based statistical approaches are developed to avoid these flaws. Such alternatives are successfully applied for thunderstorm nowcasting (Wilson and Gallant, 2000; Williams et al., 2008), and are also appearing in the context of orographic precipitation nowcasting (Panziera et al., 2010).

This study introduces an efficient computational alternative to analyse and to model orographic enhancement of precipitation from a sequence of radar images and a digital elevation model (DEM). The study considers how the terrain features such as terrain convexity and slope exposure to mesoscale flows help in explaining persistent patterns of orographic precipitation. Precipitation cells and the corresponding flow directions are extracted from radar images and attributed to the pre-computed underlying topographic variables. The orographic enhancement is defined as the ability of topography to enforce repeatability to particular precipitation patterns such as stationary cells, stable upslope ascent and localized thunderstorms. Evidences of high counts of cells repeatability reveal the topographic conditions and locations where the phenomenon is accentuated. This formulation allows characterizing precipitation enhancement using data-driven classification models. The system can be applied to simulate the localized enhancement under given flow and large scale precipitation patterns derived from nowcasting or NWP models.

The paper is organized as follows. Section 2 explains the methodology. Section 3 describes the data preparation. Its exploration is shown in Sect. 4. The computational model of orographic enhancement is explained in Sect. 5.

2 Methodology

The methodology is illustrated in the work-flow diagram in Fig. 1. It can be summarized in four main steps:

1. Compute terrain indices such as convexity and gradients at different spatial scales from the DEM.

2. Estimate the motion vector field and extract the geographical location of precipitation cells from a representative sequence of radar images of orographic precipitation events. Compute indices for slope exposure to wind direction (*flow derivative*) using the motion vector field and terrain gradients.

3. Explore the dataset using methods of clustering to find natural partitions (classes) of mesoscale flows and exposure of cells with respect to the main Alpine ridge. Select the clusters presenting potential orographic conditions (windward clusters). Within these clusters, analyse the cells' repeatability to detect the places prone to precipitation persistence and those which are not.

4. Build a statistical classification algorithm separating orographic precipitation cells from non-orographic ones in the space of features. Based on new nowcasted precipitation fields, mesoscale flows and the underlying topography, compute the susceptibility of orographic enhancement.

More details on step 1, 2 and 3 can be found in Foresti and Pozdnoukhov (2010). Preliminary results of step 4 are presented in this paper.

3 Data preparation

Radar images used for testing the methodology concern the Swiss Alps in the period from 18 to 23 August 2005. This orographic precipitation event touched in particular the northern side of the Alps (Rotach et al., 2006). Precipitation amounts exceeded 200 mm in three days with return periods above 100 yr at several weather stations (Frei, 2006). The available radar imagery has a temporal resolution of 5 min and a spatial resolution of 1×1 km^2 (Fig. 1, step 2a). It has been pre-processed to correct the vertical profiles in sheltered regions, to eliminate radar-rain gauge biases due to reduced visibility, to remove ground clutter and to account for the bright-band effect (Germann et al., 2006). The DEM used to derive the topographic information has a resolution of 250×250 m^2. The topographic features are computed at the 1×1 km^2 grid of the radar.

Data preparation passes through three main steps: the processing of the DEM, the estimation of motion vector field from subsequent radar images and the extraction of precipitation cells.

Feature extraction from DEM (Fig. 1, step 1a) was performed with Gaussian convolution filters to compute terrain convexity and terrain gradient (Fig. 1, step 1b). Features were derived at different spatial scales (degrees of smoothness) by applying convolution kernels with different bandwidths σ. More details about the extraction and the use of these features for meteorological applications can be found in Pozdnoukhov et al. (2009) and Foresti et al. (2011).

The motion vector field (Fig. 1, step 2b) is estimated from two consecutive radar images using the optical flow algorithm explained in Sun et al. (2008). Other studies consider variational techniques to the robust estimation of flow (Germann and Zawadzki, 2002). Common parameters of these algorithms allow controlling the trade-off between the precision and the spatial smoothness of the estimated field. In our approach we set the regularization parameters to have a smooth estimation of flow direction by minimizing the perturbations due to cell dissipation and growth particularly in convective situations. The flow derivative (FD) highlighting upwind slopes is computed from the terrain gradient and the motion vector field as follows:

$$\mathrm{FD}(x,t) = \nabla z(x) \cdot u(x,t) \qquad (1)$$

where $\nabla z(x)$ is the gradient vector of elevation evaluated at the (X, Y) spatial coordinates x, $u(x,t)$ is the flow vector with (u,v) components estimated at x at time t.

Several algorithms are available to detect precipitation cells from radar imagery (Lakshmanan et al., 2003; Wilson et al., 2004). In this study cells were identified by a simple method that finds the points of maxima of a smoothed precipitation field. It was done by subtracting two precipitation fields smoothed with different bandwidths σ. The resulting images describe precipitation anomalies and enable a robust selection of cells while filtering out most of clutter effects (Fig. 1, step 2b).

This processing is done on the dataset of radar images every 5 min within 6 days of precipitation (1728 images). The final dataset is composed of 28758 cells (observations) embedded in a space of 18 dimensions: [elevation | 7 convexities | 7 flow derivatives | precipitation | u,v flow components]. All input variables computed on the whole grid are stored in order to test the models under different weather situations. All data processing steps were implemented in Matlab.

4 Exploration of precipitation cells

The exploration of precipitation cells is done in two steps. First, the different flow situations (direction and strength) and the exposure of precipitation cells relative to the main Alpine ridge, described by a very large scale flow derivative, are

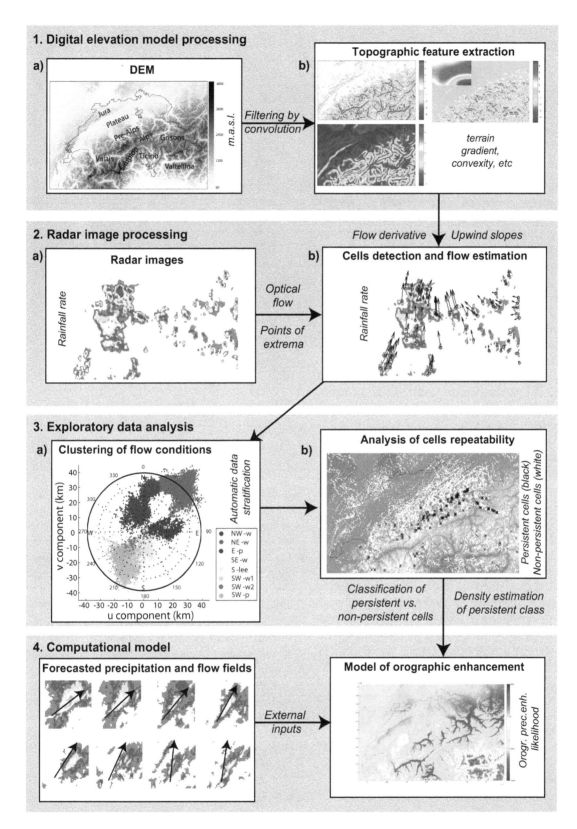

Figure 1. General scheme for data-driven modelling of orographic precipitation enhancement. External nowcasted precipitation and flow fields can be used as inputs for models of orographic precipitation enhancement, i.e. describing the likelihood of precipitation repeatability and persistence due to topography.

Table 1. Comparison of different models. AUROC and corresponding standard deviations are evaluated on 20 random splits.

Model	AUROC (st. dev.)
One-class SVM Linear	0.733 (0.008)
One-class SVM Gaussian	0.807 (0.011)
Two-class SVM Linear	0.694 (0.049)
Two-class SVM Gaussian	0.932 (0.007)

discriminated using a clustering algorithm such as k-means (Steinhaus, 1956; Hartigan and Wong, 1979) or spectral clustering (Ng et al., 2001). K-means can be used for delineating convex-shaped clusters and is nowadays used as a benchmark for weather types classification (Philipp et al., 2010). Spectral clustering (Ng et al., 2001) was used in this study because of the non spherical shape of clusters. This step is done to provide meteorological interpretability to the cells detected according to the direction of flow and their spatial location. Figure 1 step 3a plots the cells in polar coordinates according to flow direction. The different colours depict the cluster membership computed using spectral clustering in the 3-D space of (u,v) flow components an the largest scale flow derivative. Every cluster is homogeneous in terms of flow direction and relative location of cells (windward, leeward).

A detailed analysis is carried out within each cluster to recognize places which are prone to repeatability of precipitation cells. A number of counts measuring how many times a pixel is touched by a cell under similar flow conditions (same cluster) reveals a clear relationship with topography. A threshold on the counter of precipitation repeatability can be used to formulate a binary classification problem. The locations exceeding the threshold are given to the *orographic class* and the other ones are given to the *non-orographic class*. Figure 1 (step 3b) plots the geographical distribution of the two classes corresponding to the threshold value of 4. This value was empirically selected to have a sufficient number of cells representing the orographic class while keeping low the number of potential non-orographic cells falling in the orographic class. Persistent precipitation cells (orographic class) tend to concentrate in particular regions in geographical space (mainly Prealps, see Fig. 1, step 1a and step 3b) having specific topographic conditions, typically at the top of hills and on upwind slopes.

5 Computational model of orographic precipitation enhancement

The computational model of enhancement susceptibility is based on a classifier operating in the 16 dimensional space of the conditioning factors (u,v components were used only for clustering). Support vector machine (SVM, Vapnik, 1995) was selected as the classification method due to its robustness and explicit control over model's complexity. LibSVM tool-

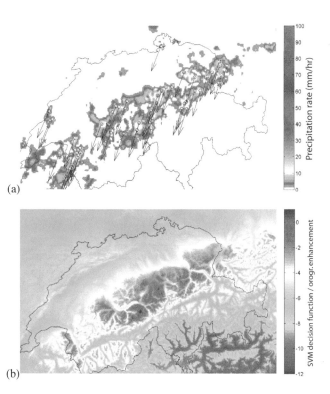

Figure 2. (a) Radar image with the detected cells and (b) the corresponding orographic enhancement characterized by the linear one-class SVM decision function.

box was used for the computations (Chang and Lin, 2001). It can be applied in a two-class and in a one-class settings (Schölkopf et al., 2001). The one-class setting considers the estimation of the support of the probability density function of the target class, the orographic cells, while discriminating the other. Both linear and non-linear class separation can be achieved by changing the kernel function encoding data similarities (dot product for a linear or Gaussian radial basis function for a non-linear separation boundary). SVM's tolerance to misclassification errors reduces the influence of the threshold value used to define the classes on the final results and allows to capture general tendencies of enhancement factors from the data.

Data were randomly split into training (50% of the data), validation (25%) and testing (25%) datasets respectively for training, model selection and assessment purposes. Table 1 shows the areas under ROC curves (AUROC, Wilks, 2005) of the test dataset after parameters selection on the validation dataset. Maximum separability is obtained with an AUROC of 1, no separability between patterns with an AUROC of 0.5. The high AUROC values for all models considered point out that orographic and non-orographic classes are separable in the high-dimensional space of topographic features. Hence, the decision function of the classifiers can be interpreted as an indicator of orographic enhancement, i.e. the ability of producing repeatability effects and persistence on precipitation.

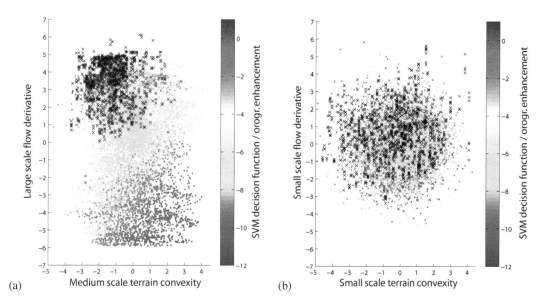

Figure 3. (a) Scatterplot of the orographic class (black crosses) in the space of features medium scale terrain convexity and large scale flow derivative. The SVM decision function is computed on the whole grid of Fig. 2b and is here displayed with the same colour scale. Orographic enhancement increases from the bottom right corner (valley bottom, leeward side of Alps) to the top left corner (hill top, windward side of Alps). (b) Same as (a) but with features smallest scale terrain convexity and smallest scale flow derivative. No patterns can be seen in this combination of features which is also neglected by the model.

Once the model is trained on a representative dataset it can be used for spatial predictions of precipitation enhancement under different flow and smooth precipitation patterns. Figure 2 shows an example of the system applied to characterize the orographic enhancement (Fig. 2b) with north-easterly flows and precipitation blocking in the north flank of the Alps (Fig. 2a). High enhancement values are found on the upwind northern side of Alps which is consistent with the blocking situation. Moreover, features due to the integration of elevation and terrain exposure can be noticed.

A key property of SVM is the ability of eliminating the irrelevant input information by weighting the different topographic and flow-related features. Thus, prediction maps are an optimal mixture of input features where the relevant ones dominate the spatial patterns and the irrelavant ones are simply filtered out. A close look to Fig. 2b indicates that patterns are likely to be produced only by a subset of the 16 features used. It suggests that terrain variables such as hills, ridges and upwind slopes need having a certain spatial scale (extension and size) to affect and be explanatory variables of precipitation persistence.

The study of features' relevance is better approached by plotting the orographic enhancement susceptibility indicator in the space of features. Figure 3 shows the same predictions of Fig. 2b but visualised respectively in a space composed of explanatory features (Fig. 3a) and in a space of irrelevant features (Fig. 3b). The SVM decision function in Fig. 3a depicts very well the membership to the orographic class constructed from the available persistent cells as a function of terrain convexity and flow exposure. On the other hand, no clear patterns can be seen in Fig. 3b.

6 Conclusions

This study introduced a generic data-driven methodology to study the orographic enhancement of precipitation. It aimed at discovering the persistent topography-related patterns of precipitation repeatability from high resolution radar images without using computationally demanding numerical models.

The extraction of precipitation cells, the estimation of mesoscale flows from radar images and the understanding of their connection to the underlying topography was the key point of the work. It allowed to reveal relevant variables for explaining patterns of orographic precipitation at different spatial scales. The exploratory analysis of the dataset with a clustering algorithm highlighted similar *weather types* in terms of mesoscale flows and exposure to the main Alpine crest (windward or leeward). Additional analyses whithin these clusters were performed to detect geographical locations prone to precipitation persistence, i.e. the places which were repeatedly touched by precipitation cells. Such places were found to be located at the top of hills and on upwind slopes. The patterns of precipitation repeatability and persistence were observed in the range of spatial scales represented by terrain features, i.e. between the micro- and the meso-gamma scales. However, only a subset of the considered scales were found to be relevant to orographic precipitation.

The evidence of separability of precipitation cells patterns motivated the construction of data-driven classification models in the high-dimensional space of conditioning variables such as topographic and flow features. The classification of cells into orographic and non-orographic, defined using a threshold on precipitation repeatability, was approached using support vector machines and provided remarkable empirical performances. The SVM decision function, which can be interpreted as a susceptibility indicator of orographic enhancement, represents how likely topographic, flow and large scale precipitation conditions produce repeatability effects on small scale precipitation patterns.

The data-driven modelling of small scale precipitation enhancement patterns in complex topography provides observational support to operational NWP including the convection-permitting models (Migliorini et al., 2011). Radar-based susceptibility maps of orographic precipitation could be used to correct the windward and leeward quantitative precipitation estimation biases present in many NWP models (Bauer et al., 2011). An important issue for the future work is to analyse larger datasets of precipitation persistence and to construct more robust predictive data-driven models which are representative of a broader set of flow and atmospheric conditions.

Acknowledgements. The research is funded by the SNSF project *GeoKernels: kernel-based methods for geo- and environmental sciences (Phase II)* (No. 200020-121835/1). A. Pozdnoukhov acknowledges the support of Science Foundation Ireland under the National Development Plan, particularly through Stokes Award and Strategic Research Cluster grant (07/SRC/I1168). We acknowledge Meteoswiss for providing the data. We thank Marco Gabella, Urs Germann, Pradeep Mandapaka and Luca Panziera of the Radar and satellites group at Meteoswiss for the interesting discussions.

Edited by: M. Dolinar
Reviewed by: M. Stoll and another anonymous referee

References

Bauer, H.-S., Weusthoff, T., Dorninger, M., Wulfmeyer, V., Schwitalla, T., Gorgas, T., Arpagaus, M., and Warrach-Sagi, K.: Predictive skill of a subset of models participating in D-PHASE in the COPS region, Q. J. Roy. Meteor. Soc., 137, 287–305, 2011.

Chang, C.-C. and Lin, C.-J.: LIBSVM: a library for support vector machines, Tech. rep., Software available at: http://www.csie.ntu.edu.tw/~cjlin/libsvm/, 2001.

Foresti, L. and Pozdnoukhov, A.: Exploration of alpine orographic precipitation patterns with radar image processing and clustering techniques, Meteorol. Appl., submitted, 2010.

Foresti, L., Tuia, D., Kanevski, M., and Pozdnoukhov, A.: Learning wind fields with multiple kernels, Stoch. Env. Res. Risk A., 25(1), 51–66, 2011.

Frei, C.: Eine Länder übergreifende Niederschlagsanalyse zum August Hochwasser 2005. Ergänzung zu Arbeitsbericht 211., Tech. rep., Arbeitsbericht MeteoSchweiz Nr. 213, 2006.

Germann, U. and Zawadzki, I.: Scale-dependence of the Predictability of Precipitation From Continental Radar Images. Part I: Methodology, Mon. Weather Rev., 130, 2859–2873, 2002.

Germann, U., Galli, G., Boscacci, M., and Bolliger, M.: Radar precipitation measurement in a mountainous region, Q. J. Roy. Meteor. Soc., 132(618), 1669–1692, 2006.

Gray, W. R. and Seed, A. W.: The characterisation of orographic rainfall, Meteorol. Appl., 7(2), 105–119, 2000.

Hartigan, J. A. and Wong, M. A.: A k-means clustering algorithm, Appl. Stat., 28, 100–108, 1979.

Lakshmanan, V., Rabin, R., and DeBrunner, V.: Multiscale storm identification and forecast, Atmos. Res., 67-68, 367–380, 2003.

Migliorini, S., Dixon, M., Bannister, R., and Ballard, S.: Ensemble prediction for nowcasting with a convection-permitting model – I: description of the system and the impact of radar-derived surface precipitation rates, Tellus A, 63(3), 468–496, doi:10.1111/j.1600-0870.2010.00503.x, 2011.

Ng, A. Y., Jordan, M. I., and Weiss, Y.: On spectral clustering: analysis and an algorithm, in: Advances on Neural Information Processing Systems, Vol. 14, 2001.

Panziera, L. and Germann, U.: The relation between airflow and orographic precipitation on the southern side of the Alps as revealed by weather radar, Q. J. Roy. Meteor. Soc., 136, 222–238, 2010.

Panziera, L., Germann, U., Gabella, M., and Mandapaka, P. V.: NORA – Nowcasting of orographic rainfall by means of analogs, Q. J. Roy. Meteor. Soc., submitted, 2010.

Philipp, A., Bartholy, J., Beck, C., Erpicum, M., Esteban, P., Fettweis, X., Huth, R., James, P., Jourdain, S., Kreienkamp, F., Krennert, T., Lykoudis, S., Michalides, S. C., Pianko-Kluczynska, K., Post, P., Álvarez, D. R., Schiemann, R., Spekat, A., and Tymvios, F. S.: Cost733cat – A database of weather and circulation type classifications, Phys. Chem. Earth, Parts A/B/C, Special Issue: Classifications of Atmospheric Circulation Patterns – Theory and Applications, 35(9–12), 360–373, 2010.

Pozdnoukhov, A., Foresti, L., and Kanevski, M.: Data-driven topo-climatic mapping with machine learning methods, Nat. Hazards, 50, 497–518, 2009.

Rotach, M., Appenzeller, C., and Albisser, P. E.: Meteoschweiz: 2006, Starkniederschlagsereignis August 2005, Tech. Rep. 211, Arbeitsberichte der MeteoSchweiz, 211, 63 pp., 22 pp., 2006.

Rotunno, R. and Houze, R. A.: Lessons on orographic precipitation from the Mesoscale Alpine Programme, Q. J. Roy. Meteor. Soc., 133, 811–830, 2007.

Schölkopf, B., Platt, J., Shawe-Taylor, J., Smola, A. J., and Williamson, R. C.: Estimating the support of a high-dimensional distribution, Neural Comput., 13, 1443–1471, 2001.

Steinhaus, H.: Sur la division des corps en parties, Bulletin de l'academie polonaise de sciences, C1 II vIV, 801–804, 1956.

Sun, D., Roth, S., and Black, M.: Learning optical flow, in: European Conference on Computer Vision, Part 3, 83–97, 2008.

Vapnik, V.: The Nature of Statistical Learning Theory, Springer-Verlag Berlin, 1995.

Wilks, D. S.: Statistical Methods in the Atmospheric Sciences, 2nd Edn., Academic Press, 2005.

Williams, J. K., Ahijevych, D. A., Kessinger, C. K., Saxen, T. R., Steiner, M., and Dettling, S.: A machine learning approach to finding weather regimes and skillful predictor combinations for short-term storm forecasting, 6th Conference on Artificial Intelligence Applications to Environmental Science, 13th Conference on Aviation, Range and Aerospace Meteorology, J1.4, 2008.

Wilson, J. P. and Gallant, J. C. (Eds.): Terrain analysis: Principles and applications, Wiley, 2000.

Wilson, J. W., Ebert, E. E., Saxen, T. R., Roberts, R. D., Mueller, C. K., Sleigh, M., Pierce, C. E., and Seed, A.: Sydney 2000 forecast demonstration project: convective storm nowcasting, Weather and Forecast., 19, 131–150, 2004.

On the role of the planetary boundary layer depth in the climate system

I. Esau[1,2] and S. Zilitinkevich[1,3,4]

[1]G.C. Rieber Climate Institute of the Nansen Environmental and Remote Sensing Center, Thormohlensgt. 47, 5006, Bergen, Norway
[2]Bjerknes Centre for Climate Research, Bergen, Norway
[3]Division of Meteorological Research, Finnish Meteorological Institute, Helsinki, Finland
[4]Division of Atmospheric Sciences and Geophysics, Department of Physics, University of Helsinki, Finland

Abstract. The planetary boundary layer (PBL) is a part of the Earth's atmosphere where turbulent fluxes dominate vertical mixing and constitute an important part of the energy balance. The PBL depth, h, is recognized as an important parameter, which controls some features of the Earth's climate and the atmospheric chemical composition. It is also known that h varies by two orders of magnitude on diurnal and seasonal time scales. This brief note highlights effects of this variability on the atmospheric near-surface climate and chemical composition. We interpret heat capacity parameter of a Budyko-type energy balance model in terms of quasi-equilibrium h. The analysis shows that it is the shallowest, stably-stratified PBL with the smallest h that should be of particular concern for climate modelling. The reciprocal dependence between the PBL depth and temperature (concentrations) is discussed. In particular, the analysis suggests that the climate characteristics during stably stratified PBL episodes should be significantly more sensitive to perturbations of the Earth's energy balance as well as emission rates. On this platform, h from ERA-40 reanalysis data, the CHAMP satellite product and the DATABASE64 data were compared. DATABASE64 was used to assess the Troen-Mahrt method to determine h through available meteorological profile observations. As it has been found before, the shallow PBL requires better parameterization and better retrieval algorithms. The study demonstrated that ERA-40 and CHAMP data are biased toward deeper h in the shallow polar PBL. This, coupled with the scarcity of in-situ observations might mislead the attribution of the origins of the Arctic climate change mechanisms.

1 Introduction

The lowermost atmospheric layer where the vertical turbulent exchange is significant is known as the planetary boundary layer (PBL). The importance of the PBL for the Earth's climate system has been recognized since pioneering work of Manabe and Strickler (1964). However, there are just a few studies (e.g. Knight et al., 2007), which excurse beyond the current narrow focus on PBL parameterizations in climate models. Using a classification and regression tree approach, Knight et al. (2007) demonstrated with 57 067 climate model HadAM3 runs that 80% of variation in climate sensitivity to $2 \times CO_2$ is associated with variation of a small subset of parameters mostly related to the convection processes. The most influential parameter (32% of the total variability) appeared to be the entrainment rate, A, which defines the top down (counter-gradient) turbulent flux at the PBL upper boundary. The PBL mass budget leads to the direct proportionality between the PBL depth, h, and the entrainment rate (Medeiros et al., 2005). Knight et al. (2007) reported the highest climate sensitivity for the lowest A, found in the shallowest PBL. The reciprocal relationship between the climate sensitivity and the PBL depth suggests larger climate variability in stably stratified boundary layers (Esau, 2008; Zilitinkevich and Esau, 2010), where h is small, typically less than 200 m in nocturnal PBL as to be compared to its daytime counterpart with $h > 1000$ m (e.g. Medeiros et al., 2005). So far, very little attention has been paid to the stably stratified PBL. Its parameterizations are unreliable (Beare et al., 2006; Cuxart et al., 2006) and often overestimate h in simulations (Mauritsen et al., 2007). Corresponding damping of model sensitivity has been demonstrated at high-latitudes (Beesly et al., 2000; Byrkjedal et al., 2008). Nevertheless, at present, the primary attention of climatologists is still focused on the daytime PBL (Stone and Weaver, 2003; Walters et al., 2007). This study consists of two distinct sections. Section 2 looks at the bulk PBL effect and the PBL depth in a simple energy balance model. Section 3 addresses challenges of PBL depth diagnostics.

2 Bulk planetary boundary layer effect on the climate system

In spite of remarkable progress in climate modelling and observation systems, a simple zero-dimensional energy balance climate model remains instrumental in understanding of the climate processes and mechanisms (North et al., 1981). In particular, utility of these models is in suggesting directions for statistical analysis of more comprehensive data from observations and modelling in order to separate certain physical effects. Following (Budyko, 1969; North et al., 1981; Esau, 2008; Zilitinkevich and Esau, 2010), we consider a Budyko-type energy balance model. The model reads

$$C\frac{dT}{dt} \propto F_T - F_{T0}, \tag{1}$$

where T is temperature, F_T is the divergence of the temperature flux, and C is the heat capacity of the system. On the Earth, positive and negative F_T alternate on different time scales, primarily on diurnal and annual time scales, to make climate, i.e. the state where the averaged over many years $dT/dt \to 0$. Addition of the averaged equilibrium flux, F_{T0}, for which $dT/dt \to 0$, allows studies of a transient climate change in deviations from a unknown equilibrium state. Traditionally, the attention of the climate research community is focused on investigation of perturbations caused by F_{T0} response on the shift of the radiation heat balance. North et al. (1981) review and the very recent Zaliapin and Ghil (2010) insightful analysis are just two examples of such a work. Contrary, the system heat capacity, C, has not received much attention. In a system where the role of turbulent mixing is significant, one can simplify as $C \propto h$. Here, we assumed that complexity of the turbulent mixing processes could be parameterized through a single integral parameter, namely, the PBL depth h. The interested reader can find reservations and restrictions of this approach as well as supporting data in Esau (2008) and Zilitinkevich and Esau (2010).

As it follows from the traditional climatological convention, T should be understood as the mean (global and seasonal averaged) surface air temperature. This convention immediately raises problems of aggregation of the system heat capacity and the temperature fluxes. We will not excurse into those problems making a silent assumption that both quantities are aggregated properly. Moreover, we assume that our model in Eq. (1) does not have sub-scale temporal variability. The temperature change represents the direct transition

of the system from the unperturbed to a perturbed state. One should remember that these energy balance models are not to quantify behavior of the real Earth's climate system, which is incomparably more complex, but to suggest approaches to data analysis.

Thus, with respect to the PBL effects, the model in Eq. (1) reads

$$\frac{dT}{dt} = c_{T0}\frac{F_T - F_{T0}}{h}. \tag{2}$$

Here, c_{T0} is a non-dimensional proportionality coefficient, h is dynamic quantity ultimately depending on F_{T0} and therefore on T, which suggest a possible non-linear PBL feedback in this kind of models. Esau (2008) and Zilitinkevich and Esau (2010) provided more materials to elaborate on the PBL feedback. We, however, will focus only on the bulk effects related to large variability of h.

Equation (2) immediately reveals several falsifiable propositions: (a) the temperature response to a given flux perturbation should have larger magnitude in the shallower PBL where h is small; (b) the temperature variability should be larger in the shallow PBL; and (c) the temperature change is faster in the shallow PBL. One should observe that the temperature has probably stronger links with radiation processes and atmospheric large-scale dynamics hidden in F_{T0} than with the vertical turbulent mixing hidden in h. Manabe and Strikler (1964) and following works with the radiation-convective models (Ramanathan and Coakley, 1978; Randall et al., 1996) have demonstrated that this is not the case for the global scale climate as such. However, it could be the case for small perturbations (e.g. due to doubling of $CO2$ concentration) and on local scales. Moreover, only the local scale data will be available for direct analysis of the PBL effects as the nature of aggregation process remains unspecified.

One of possible ways to circumvent, at least partially, those difficulties is to recast Eq. (2) in terms of a scalar concentration, Q. It reads

$$\frac{dQ}{dt} = c_{Q0}\frac{F_Q - F_{Q0}}{h}, \tag{3}$$

where $c_{Q0} \neq c_{T0}$ is a non-dimensional proportionality coefficient, Q is the scalar concentration and F_{Q0} is the divergence of the scalar flux. If Q represents a long-lived scalar with its emission source predominantly within the PBL then, over sufficiently large homogeneous area, F_{Q0} is mostly controlled by the turbulent mixing. In this case, the bulk PBL effect could be separated in almost pure, laboratory conditions.

This study is not aimed to publish comprehensive analysis of the PBL effects in the Earth's system data. Therefore, it will be only demonstrated that the existing data are consistent with the proposed effects (a)–(c) both on global and local scales. First, we envisage the way to approach the temperature analysis. The PBL depth is the key parameter in Eq. (2). The variability of the PBL depth could be simplified as: $h_C \sim O(10^3\,\mathrm{m})$ with the typical value $\overline{h_C} = 1500\,\mathrm{m}$,

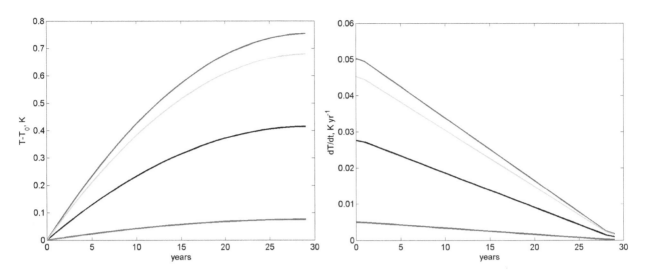

Figure 1. Response (**a**) of the temperature after Eq. (1) and its change rate (**b**) on an instant perturbation of the temperature flux by the observed green-house gas forcing. The constant $c_{T0} = 1000$ has been chosen in the way to bring the mean temperature change close to the observed one over the last 30 years. The black line denotes the mean temperature $0.5\,(T_{\max} + T_{\min})$; the blue line – the minimum temperature $T_{\min}$; the red line – the maximum temperature $T_{\max}$; the green line – the diurnal temperature range change $(T_{\min} - T_{\max})$. The initial state has been subtracted for all quantities.

– a characteristic depth of the convective, mostly daytime PBL; and $h_S \sim O(10^2\,\text{m})$ with the typical value $\overline{h_S} = 150\,\text{m}$, – a characteristic depth of the stably-stratified, mostly night-time PBL. These two distinct types of the PBL could be segregated in data and the corresponding temperature changes could be calculated. Since, in the most of cases, the turbulent convection is a dynamical reaction of the atmosphere on the positive surface heat balance, the convective PBL should be characterized by temperature maximums. Contrary, the stably stratified PBL forms during periods with the negative heat balance. Hence it should be characterized by temperature minimums.

Several studies of the diurnal temperature maximum, $T_{\max}$, and minimum, $T_{\min}$, have been published to date (e.g. Hansen et al., 1995; Braganza et al., 2004; Vose et al., 2005). We consider here how the energy balance model in Eq. (2) will react on the gradual change of the forcing. The globally averaged green-house gas temperature forcing is $\Delta F_{T0} \sim 2.6 \times 10^{-5}\,\text{K m s}^{-1}\,\text{yr}^{-1}$ for 30 year between 1979 and 2008 (Annual Greenhouse Gas Index, NOAA Earth System Research Laboratory, Global Monitoring Division, http://iasoa. org/iasoa/index.php). We impose a new climate equilibrium with $F_{T1} = F_{T0} + 30\Delta F_{T0}$ and integrate Eq. (2) to find the minimum and maximum temperature adjustment in the deep and shallow PBLs correspondingly. To do this, we plot in Fig. 1 the following illustrative solution

$$T(t) - T(t=0) = c_{T0}\frac{F_{T1} - F_T(t)}{h} = c_{T0}\frac{30\Delta F_{T0} - t\Delta F_{T0}}{h}. \quad (4)$$

As one can see, $\Delta T_{\max}$ increases slower and changes on much smaller value than $\Delta T_{\min}$ does. Therefore, the diurnal temperature range (DTR) reduces. Its reduction accumulates

with time but the strongest reduction would be seen during periods of the strongest temperature flux forcing. Interesting that in warmer climate, it is rather unlikely to observe decrease of $\overline{h_C}$ or $\overline{h_S}$ since the latter is already too small to show significant relative variations and the former cannot decrease in response on increasing temperature flux. There will be probably less cases and shorter periods with $\overline{h_S}$, which does not help to increase DTR either. Thus, the differential extreme temperature change and the decrease of the DTR over long time should be considered as the robust signature of the climate warming caused by the radiation balance shift due to the change of the atmospheric composition. In simple words, the difference in the PBL depth during day and night time leads to greatly suppressed response in the daytime temperature relative to that in the night time.

Rudimentary and idealized estimations given above are however in surprisingly good agreement with available analysis of observations. Braganza et al. (2004) give the rates $\Delta T_{\max} = 0.1\,\text{K dec}^{-1}$ and $\Delta T_{\min} = 0.2\,\text{K dec}^{-1}$ for the 50 years' period between 1950 and 2000, which leads to $(T - T_0)_{\max} = 0.3\,\text{K}$ and $(T - T_0)_{\min} = 0.6\,\text{K}$ over 30 years. However, Vose et al. (2005) and Hansen et al. (1995) found $\Delta T_{\max} \approx \Delta T_{\min} \approx 0.29\,\text{K dec}^{-1}$ for a shorter period of time. The energy balance model suggests that there should be continues decrease of the DTR, albeit probably at slower rate, until new equilibrium has been reached. One should remind however that 70% of the Earth's surface is covered by ocean where the amplitude of the diurnal cycle in the PBL is significantly reduced.

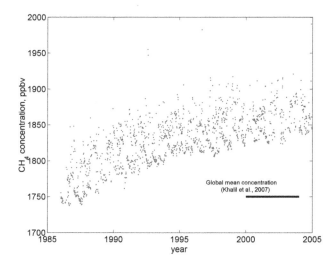

Figure 2. Daily averaged methane concentration obtained from NOAA/CMDL data at Barrow: red dots – summertime (June–July) data; blue dots – wintertime (December–January) data.

The differential temperature change is of primarily interest for climate research. The interest in the concentration variability comes from many applications dealing also with shorter than decadal time scales. The bulk PBL effect on seasonal concentration of methane, CH_4, is shown in Fig. 2. This Figure reveals very powerful bulk PBL effect. The near surface concentration of CH_4 is systematically higher (by about 100 parts per billion in volume, ppbv) than the concentration of CH_4 in the free atmosphere (Khalil et al., 2007). This is expected as the primary CH_4 emission is originated from the surface (Walter et al., 2006). Since about 1998–1999, the average near-surface seasonal concentrations do not increase which is consistent with the global free atmosphere records. Thus, one can conclude that the local emission of CH_4 remains nearly constant over the period of observations or it varies as a proportion of the free atmosphere methane concentration, which is highly unlikely. Hence, both winter and summertime CH_4 concentrations are determined by the local emission of the gas. Mastepanov et al. (2008) found significant methane release on the onset of the freezing season, which is a border between somewhat significant methane emission during the short polar summer (90 days or so) and much lower emission during the long polar winter (~270 days). Despite lower emission rate in wintertime, the CH_4 concentration remains significantly higher. It exhibits large (~50 ppbv) variations on a week scale, which suggest that the individual release events could be mixed up into the free atmosphere during a week or so. At the same time, CH_4 concentration always remains higher than that in summertime. These facts can be explained with Eq. (3). Both larger CH_4 concentration and larger variability are understood as the effect of significant reduction of the PBL depth in response on the negative radiation balance and corresponding increase of the atmospheric stability. The convective mo-

tions are damped that both confines CH_4 in the shallow layer of $\overline{h_S} \sim 100\,m$ and prevents its exchange with the free atmosphere due to build up of a strong radiation temperature inversion. ERA-40 (Uppala et al., 2005) data (a data selection within $2.5°$ by $2.5°$ rectangle centred at Barrow) revealed that $\overline{h_C} - \overline{h_S}$ is $\sim 300\,m$. These data allow for rough estimation of the CH_4 emission rate using only meteorological and concentration measurements. The mass of CH_4 (per unit area), M_Q, emitted into the PBL should be equal to the mass of CH_4 mixed into the free atmosphere during about 10 days. It can be calculated as

$$M_Q = \frac{1}{2}(Q - Q_0)h. \tag{5}$$

Now, we substitute for summertime $Q = 1820\,ppbv$, $Q_0 = 1760\,ppbv$ and $\overline{h_C} = 400\,m$. It gives the emission rate $M_Q \sim 1.2 \times 10^{-5}\,kg\,m^{-2}\,day^{-1}$, which is fairly consistent with $\sim 1.0 \times 10^{-5}\,kg\,m^{-2}\,day^{-1}$ obtained by Walter et al. (2007) in direct measurements. In wintertime, the numbers are as follows, $Q = 1850\,ppbv$, $Q_0 = 1760\,ppbv$ and $\overline{h_C} = 100\,m$. It gives the emission rate $M_Q \sim 4.5 \times 10^{-7}\,kg\,m^{-2}\,day^{-1}$, which is more than order of magnitude smaller than that in summertime. Thus, shallower PBL in wintertime could maintain the higher CH_4 concentration in spite of the reduced emission.

3 Boundary layer depth diagnosis using observations, models and satellite products

The PBL effects on climate are difficult to study because h is not regularly observed quantity. It is diagnosed both in observations and simulations. Unfortunately, models reveal systematic overestimation and generally poor correlation between the simulated and observed h in both convective and stably stratified PBL (Cuxart et al., 2006; Steeneveld et al., 2008). From Eq. (2) however follows that the bias in h could lead to biased climatology. In order to estimate this bias, we used the ERA-40 data (Uppala et al., 2005) for the Barrow grid point as described above. Figure 3 shows the PBL depth estimated in the reanalysis and the satellite product CHAMP (von Engeln, 2005) as compared with the theoretical evaluation using corresponding ERA-40 external parameters (surface heat flux, atmospheric stability, and the wind speed) and the analytical relationship proposed and calibrated in Zilitinkevich et al. (2007). Figure 3 suggests a tendency of the ECMWF model (the model used for the ERA-40 reanalysis) to over-predict h in stably stratified PBL revealed also in Mauritsen et al. (2007). The ERA-40 (blue) curve deviation in Fig. 3b is rather similar to the one found by Mauritsen et al. (2007) in direct intercomparison with the DATABASE64 cases. The over-prediction by the CHAMP retrieval algorithm could be possibly traced to that in ECMWF model, which is used to initialize the iteration retrieval algorithm (Marquardt et al., 2002). The good correspondence between CHAMP and ERA-40 data in tropical latitudes (von Engeln, 2005) is not confirmed for the high latitudes. Statistically,

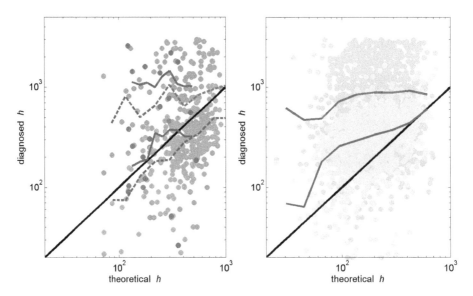

Figure 3. The daily averaged stably stratified quasi-steady state PBL depth, h . Days with convection and Monin-Obukhov length changes larger than $\Delta L = 50$ m per day are excluded. Following Zilitinkevich et al. (2007), data were separated in (**a**) nocturnal (solid curves; light dots) and conventionally neutral (dashed curves; dark dots) PBLs; (**b**) long-lived stably stratified PBL. Bluish dots and blue curves are the ERA-40 data. Reddish dots and red curves are the CHAMP data. The curves are the bin-averaged values of the corresponding data.

the ERA-40 data are in good agreement with the present understanding and turbulence-resolving simulation of the nocturnal and near-neutral PBLs. Parameterizations for those layers have been thoroughly calibrated for the purpose of the mid-latitude weather prediction with the model. At the same time, the long lived stably stratified PBL remains a challenge for the model.

The over-prediction of the shallow PBL and the underprediction of the deep PBL remain a challenging problem for other models as well. For instance Han et al. (2008) compared 5 PBL schemes in the Weather Research and Forecasting (WRF) model with observations in the Hong Kong area. Totally 145 samplings of h were collected in 22 flights in March 2001. Both geographical and diurnal cycle patterns of h showed only moderate correlation (correlation coefficients vary from 0.65 to 0.7) to the observations. Typical nocturnal biases were found to be of +300 m or about 100% of the observed h values. Typical daytime biases were found to be −300 m or about 30% of the observed h values.

The diagnosis of the PBL depth in simulations (including ERA-40 reanalysis) and observations is usually based on the bulk Richardson number method by Troen and Mahrt (1986), which defines

$$h: \quad Ri(z) = Ri_{cr}, \quad \text{where} \quad Ri = \frac{g}{\theta_0} \frac{\theta_h - \theta_s}{u_h^2} z, \qquad (6)$$

θ_s is the surface potential temperature; θ_h is the potential temperature at h; u_h is the wind speed at h; $g/\theta_0 = 0.03$ m s^{-2} K^{-1} is the buoyancy parameter; and z is the height above surface. The method could be sensitive to a calibration of the critical value Ri_{cr}. In meteorological practice, Ri_{cr} are taken in

the range from 0.15 to 1.0 with often quoted values of 0.25 and 0.5 (Serafin and Zardi, 2005; Jeričević and Grisogono, 2006). We investigated the variability of h as function of Ri_{cr} in the stably stratified PBL collected in the large-eddy simulation database (The DATABASE64 could be downloaded from ftp://ftp.nersc.no/igor/NEW%20DATABASE64/; Esau and Zilitinkevich, 2006). Since the large-eddy simulations resolve the PBL turbulence, the DATABASE64 gives the opportunity to determine h by different methods and to compare the results of directly determined PBL depth, i.e. using turbulent flux profiles. Figure 4 shows the relative error, $(h_{TM} - h_{LES})/h_{LES}$, as a function of Ri_{cr} for the conventionally neutral, nocturnal stable and long-lived stable PBL averaged over all corresponding runs in DATABASE64. Here, h_{TM} is the PBL depth defined by Eq. (6), h_{LES} is the PBL depth defined as a level where the vertical momentum flux is equal to 5% of its surface value. The minimum error is found on average for Ri_{cr} in the range 0.15 to 0.25. Excluding the nocturnal PBL, we conclude that the Troen–Mahrt method is robust with optimal $Ri_{cr} = 0.2$ and a negligible relative error. In the weakly stratified nocturnal PBL, h is poorly defined. We also repeated these calculations with the Arctic Ocean Experiment data (Tjernstroem et al., 2005) that provide independent estimations of h based on radiometer data and humidity, wind and temperature profiles. Those calculations provided very similar dependence to ones shown in Fig. 4.

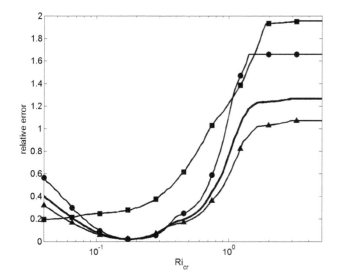

Figure 4. The relative error $(h_{TM} - h_{LES})/h_{LES}$, as functions of Ri_{cr} for the nocturnal stable (squares), conventionally neutral (circles) and long-lived stable (triangles) PBL averaged over all corresponding runs in DATABASE64. The bold line without symbols is the mean line for the later two types of the PBL.

4 Conclusions

This work is motivated by the fact that the planetary boundary layer depth, h, is important for the Earth's climate. The first step to understand the bulk PBL effect on the Earth's climate is to consider the energy-balance model. The model leads to reciprocal dependence between the temperature/concentration changes and h. The dependence can be summarized as: (a) the temperature response to a given flux perturbation has larger magnitude in the shallower PBL where h is small; (b) the temperature variability should be larger in the shallow PBL; and (c) the temperature change is faster in the shallow PBL. These points were illustrated using the temperature response on the green-house gas forcing since 1979 and the seasonal variability of the methane concentration at point Barrow (Alaska, USA). We found reasonable agreement in the observed trends and variability between the methane concentrations, temperature anomalies and the PBL depth with numbers found in literature. We demonstrated that the model parameterizations (ERA-40) and the satellite products (CHAMP) over-predict h in shallow long-lived PBL typical in high-latitudes. This bias could and probably does cause climate model discrepancies in polar areas as those reported by Beesly et al. (2000). This bias and the scarce in situ observations in the region makes it difficult to ascribe the polar climate change pattern.

Acknowledgements. The research leading to these results has received funding from the EC FP7/2007-2011 programme under grant agreement no. 212520, the EC ERC programme grant PBL-PMES and the Norwegian Research Council grant PBL-FEEDBACK.

Edited by: A. M. Sempreviva
Reviewed by: three anonymous referees

References

Beare, R. J., MacVean, M. K., Holtslag, A. A. M., Cuxart, J., Esau, I., Golaz, J.-C., Jimenez, M. A., Khairoutdinov, M., Kosovic, B., Lewellen, D., Lund, T. S., Lundquist, J. K., McCabe, A., Moene, A. F., Noh, Y., Raasch, S., and Sullivan, P.: An intercomparison of large-eddy simulations of the stable boundary layer, Bound.-Lay. Meteorol., 118(2), 2, 247–272, 2006.

Beesley, J. A., Bretherton, C. S., Jakob, C., Andreas, E. L., Intrieri, J. M., and Uttal, T. A.: A comparison of cloud and boundary layer variables in the ECMWF forecast model with observations at Surface Heat Budget of the Arctic Ocean (SHEBA) ice camp, J. Geophys. Res., 105(D10), 12337–12350, 2000.

Braganza, K., Karoly, D. J., and Arblaster, J.: Diurnal temperature range as an index of global climate change during the twentieth century, Geophys. Res. Lett., 31, L13217, doi:10.1029/2004GL019998, 2004.

Byrkjedal, Ø., Esau, I., and Kvamst, N.-G.: Sensitivity of simulated wintertime Arctic atmosphere to vertical resolution in the ARPEGE/IFS model, Clim. Dynam., 30(1–2), 687–701, doi:10.1007/s00382-007-0316-z, 2008.

Cuxart, J., Holtslag, A., Beare, R., Bazile, E., Beljaars, A., Cheng, A., Conangla, L., Ek, M., Freedman, F., Hamdi, R., Kerstein, A., Kitagawa, H., Lenderink, G., Lewellen, D., Mailhot, J., Mauritsen, T., Perov, V., Schayes, G., Steeneveld, G.-J., Svensson, G., Taylor, P., Weng, W., Wunsch, S., and Xu, K.-M.: Single-Column Model Intercomparison for a Stably Stratified Atmospheric Boundary Layer, Bound.-Lay. Meteorol., 118(2), 273–303, doi:10.1007/s10546-005-3780-1, 2006.

von Engeln, A., Teixeira, J., Wickert, J., and Buehler, S. A.: Using CHAMP radio occultation data to determine the top altitude of the Planetary Boundary Layer, Geophys. Res. Lett., 32(6), L06815, doi:10.1029/2004GL022168, 2005.

Esau, I: Formulation of the Planetary Boundary Layer Feedback in the Earth's Climate System, Computational Technologies, 13, special issue 3, 95–103, 2008.

Esau, I. N. and Zilitinkevich, S. S.: Universal dependences between turbulent and mean flow parameters instably and neutrally stratified Planetary Boundary Layers, Nonlin. Processes Geophys., 13, 135–144, doi:10.5194/npg-13-135-2006, 2006.

Han, Z., Ueda, H., and Ana, J.: Evaluation and intercomparison of meteorological predictions by five MM5-PBL parameterizations in combination with three land-surface models, Atmos. Environ., 42, 233–249, 2008.

Hansen, J., Sato, M., and Ruedy, R.: Long-term changes of the diurnal temperature cycle: implications about mechanisms of global climate change, Atmos. Res., 37, 175–209, 1995.

Jeričević, A. and Grisogono, B.: The critical bulk Richardson number in urban areas verification and application in a numerical weather prediction model, Tellus, 58A, 19–27, 2006.

Khalil, M. A. K., Butenhoff, C. L., and Rasmussen, R. A.: Atmospheric methane: Trends and cycles of sources and sinks, Environ. Sci. Technol., 41(7), 2131–2137, doi:10.1021/es061791t, 2007.

Knight, C. G., Knight, S. H. E., Massey, N., Aina, T., Christensen, C., Frame, D. J., Kettleborough, J. A., Martin, A., Pascoe, S., Sanderson, B., Stainforth, D. A., and Allen, M. R.: Association of parameter, software, and hardware variation with large-scale behaviour across 57,000 climate models, PNAS, 104(30), 12259–12264, 2007.

Manabe, S. and Strickler, R. F.: Thermal equilibrium of the atmosphere with a convective adjustment, J. Atmos. Sci., 21, 361–385, 1964.

Marquardt, C., Beyerle, G., Healy, S. B., Schmidt, T., Wickert, J., Neumayer, H., König, R., and Reigber, Ch.: Variational Retrieval of Champ Radio Occultation Data, European Geophysical Society XXVII General Assembly, Nice, 21–26 April, abstract N6004, 2002.

Medeiros, B., Hall, A., and Stevens, B.: What Controls the Mean Depth of the PBL?, J. Climate, 18, 3157–3172, 2005.

Mastepanov, M., Sigsgaard, C., Dlugokencky, E. Houweling, S., Strom, L., Tamstorf, M. P., and Christensen, T. R.: Large tundra methane burst during onset of freezing, Nature, 456, 628–630, 2008.

Mauritsen, T., Svensson, G., Zilitinkevich, S. S., Esau, I., Enger, L., and Grisogono, B.: A total turbulent energy closure model for neutral and stably stratified atmospheric boundary layers, J. Atmos. Sci., 64(11), 4117–4130, 2007.

North, G. R., Cahalan, R. F., and Coakley Jr., J. A.: Energy balance climate models, Rev. Geophys. Space Phys., 19(1), 91–121, 1981.

Ramanathan, V. and Coakley Jr., J. A.: Climate modelling through radiative-convective models, Rev. Geophys. Space Phys., 16, 465–489, 1978.

Randall, D. A., Xu, K.-M., Somerville, R. J. C., and Iacobellis, S.: Single column models and cloud ensemble models as links between observations and climate models, J. Climate, 9, 1683–1697, 1996.

Serafin, S. and Zardi, D.: Critical evaluation and proposed refinement of the Troen and Marht (1986) boundary layer model, ICAM/MAP conference, 2005.

Steeneveld, G. J., Mauritsen, T., de Bruijn, E. I. F., Vila-Guerau de Arellano, J., Svensson, G., and Holtslag, A. A. M.: Evaluation of limited area models for the representation of the diurnal cycle and contrasting nights in CASES99, J. Appl. Meteorol. Clim., 47, 869–887, 2008.

Stone, D. A. and Weaver, A. J.: Factors contributing to diurnal temperature range trends in twentieth and twenty-first century simulations of the CCCma coupled model, Clim. Dynam., 20(5), 435–445, 2003.

Uppala, S. M., Kållberg, P. W., Simmons, A. J., et al.: The ERA-40 re-analysis, Q. J. Roy. Meteor. Soc., 131(612), 2961–3012, doi:10.1256/qj.04.176, 2005.

Tjernström, M.: The Summer Arctic Boundary Layer during the Arctic Ocean Experiment 2001 (AOE-2001), Bound.-Lay. Meteorol., 117, 5–36, 2005.

Troen, I. and Mahrt, L.: A simple model of the atmospheric boundary layer; sensitivity to surface evaporation, Bound.-Lay. Meteorol., 37, 129–148, 1986.

Vose, R. S., Easterling, D. R., and Gleason, B.: Maximum and minimum temperature trends for the globe: An update through 2004, Geophys. Res. Lett., 32, L23822, doi:10.1029/2005GL024379, 2005.

Walter, K. M., Zimov, S. A., Chanton, J. P., Verbyla, D., and Chapin III, F. S.: Methane bubbling from Siberian thaw lakes as a positive feedback to climate warming, Nature, 443, 71–75, 2006.

Walter, K. M., Smith, L. C., and Chapin III, F. S.: Methane bubbling from northern lakes: present and future contributions to the global methane budget, Phil. Trans. R. Soc. A, 365, 1657–1676, 2007.

Walters, J. T., McNider, R. T., Shi, X., Norris, W. B., and Christy, J. R.: Positive surface temperature feedback in the stable nocturnal boundary layer, Geophys. Res. Lett., 34, L12709, doi:10.1029/2007GL029505, 2007.

Zaliapin, I. and Ghil, M.: Another look at climate sensitivity, arXiv:1003.0253v1, 2010.

Zilitinkevich, S., Esau, I., and Baklanov, A.: Further comments on the equilibrium height of neutral and stable planetary boundary layers, Q. J. Roy. Meteor. Soc., 133, 265–271, 2007.

Zilitinkevich, S. S. and Esau, I.: Planetary boundary layer feedbacks in climate system and triggering global warming in the night, in winter and at high latitudes, Geography, Environment and Sustainability, 1(2), 20–34, 2009.

5

Extreme summer temperatures in Western Europe

C. Simolo[1], **M. Brunetti**[1], **M. Maugeri**[2,1], **and T. Nanni**[1]

[1]Institute of Atmospheric Sciences and Climate, National Research Council (ISAC-CNR), Bologna, Italy
[2]Dipartimento di Fisica, Università degli Studi di Milano, Italy

Correspondence to: C. Simolo (c.simolo@isac.cnr.it)

Abstract. We discuss the evolution of summer temperature extremes over Western Europe during 1961–2004 in the context of current climate warming. Using a parametric approach, we investigate the role of properties and changes in probability density functions of daily temperatures in modifying the frequency of severe, isolated events. In this perspective, the recent intensification of extremely warm events over Europe turns out to be well consistent with a pure, nonuniform shift of mean values, with no room for conjectures about increasing temperature variability.

1 Introduction

European temperatures have undergone a considerable rise in the recent past, accompanied by marked changes in the frequency and magnitude of warm and cold extremes (Alexander et al., 2006; Moberg et al., 2006). The way such changes are controlled by changes in mean climatic conditions still needs to be understood. Exceptionally warm events, for instance, are generally thought to be triggered by increasing temperature variability on the daily or annual timescale (Schär et al., 2004; Della-Marta et al., 2007). Likewise, modifications in the distributional shape of daily temperatures have been inferred from discrepancies between the observed behavior of moderate (so-called *soft*) extremes and that expected under a pure shift of mean temperatures (Klein Tank and Können, 2003; Klein Tank et al., 2005). Although such a climate shift seems inadequate to explain the recent intensification of unusually warm events, there is no clear observational evidence for changes in the statistical properties beyond the mean (e.g. Scherrer et al., 2005; Simolo et al., 2010, 2011).

A key-point is therefore to investigate the relationships between the behavior of severe, isolated events and the average distributional properties of daily temperatures. Here we provide an appropriate theoretical framework to determine the exact nature of these relationships, that is, for translating basic properties and long-term trends in the moments of probability density functions (PDFs) of daily temperatures into changes in the frequency of warm and cold extremes.

This formalism is here used to examine the summer evolution of daily temperatures over a Western European subdomain, where the past decades have seen a strong warming that caused well-known record-breaking events, and is projected to be even more severe in the near future (e.g. Fischer and Schär, 2009; Hirschi et al., 2011). Using this case as a baseline, we show how the observed evolution of soft extremes can be understood in the light of a pure shift of mean temperatures, with no need for invoking more complicated changes in the second and higher moments. For further details see Simolo et al. (2011).

A brief description of the data used and their processing is given in Sect. 2. A model for time-evolving daily temperature PDFs and soft extremes is discussed in Sect. 3. Main results are then illustrated in Sect. 4 by direct comparison with observations. Conclusions are finally drawn in Sect. 5.

2 Observations

As discussed in detail in Simolo et al. (2011), analyses of changes in European temperature extremes during the past 50 yr are based on maximum (TX) and minimum (TN) daily records from the European Climate Assessment (ECA) dataset (Klein Tank et al., 2002), selected on the basis of strict quality and homogeneity criteria. Station series, re-expressed as anomalies relative to the base period 1961–1990, were gathered in three European sub-domains using a Principal Component Analysis, i.e. a southern area from the Alpine to the Carpathian Chain (AC), a Northern Sea (NS) and an

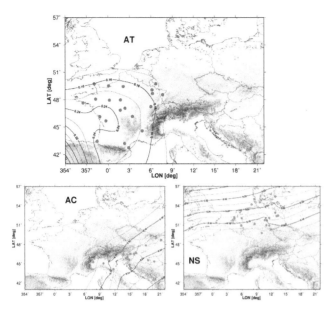

Figure 1. European areas (AT, AC and NS, see text) defined by Principal Component Analysis. Red dots in the upper panel denote locations of the 21 stations retained in the AT sub-domain. Loadings of Varimax rotated empirical orthogonal functions are also shown.

Atlantic (AT) area (see Fig. 1). Unweighted spatial averages were then derived for each sub-domain and for both TX and TN anomalies. Data availability and quality allowed to cover 1961–2007 over AC and NS and 1961–2004 over AT. Here we focus on summer (JJA) temperature evolution across the AT sub-domain, and refer to Simolo et al. (2011) for a comprehensive discussion. The observed evolution of soft extremes, i.e. those events confined to the outermost few percent tails of daily PDFs, is conventionally measured as day-counts per year (or season) exceeding/not exceeding fixed thresholds. We consider percentile-based indicators (Alexander et al., 2006) for gradually smaller fractions of the cold and the warm tail in TX PDFs, i.e. TXN, with $N = 10, 5, 1$ and $N = 90, 95, 99$ respectively, and similar for TN.

3 PDFs and soft extremes

Potential variations in the second and higher moments of daily temperature PDFs, as already noted, are usually deduced indirectly from anomalous behaviors of distribution tails. Here, reversing the point of view, we determine the time evolution of cold and warm extremes from observed properties and changes in daily temperature distributions. A theoretical model is used for describing time-evolving PDFs and their basic features (Sect. 3.1). The ensuing changes in soft extremes can be determined within this framework in terms of time-dependent exceedance probabilities (EPs), i.e. total probabilities for events exceeding/not exceeding fixed thresholds (Sect. 3.2). As an example, the whole formalism

is adjusted for the TX and TN summer evolution over the AT region, as will be discussed in Sect. 4.

3.1 A time-evolving skewed density (TESD) model

The most prominent changes in European temperatures concern the increase of the mean, whereas no significant trends during the recent past have been clearly detected in the second and higher moments, except only for weak seasonal effects (e.g. Scherrer et al., 2005; Moberg et al., 2006). This issue has been addressed in Simolo et al. (2011) by a moment-based analysis of the empirical PDFs, derived from single-year sub-samples of TX and TN anomalies for the three European areas over their respective periods (see Sect. 2). In all cases the time series of second and higher moments did not reveal significant trends, but random fluctuations only (Simolo et al., 2011). The strongest warming was detected over the AT region, with a marked summer trend in both TX and TN mean values, which is only approximately linear in time and best fitted by a second-degree polynomial (see Fig. 2a). Similar to the other European areas, either on the annual or seasonal timescale, the AT summer PDFs were found to be essentially stationary in shape, though the shape itself shows departures from normality that are related to the degree of asymmetry.

Modeling observed PDFs therefore requires a skewed extension of the normal density function, i.e.

$$f(z) = 2\phi(z)\Phi(\alpha z), \quad \text{where}$$
$$\phi(z) = \frac{1}{\sqrt{2\pi}}\exp\left(-\frac{z^2}{2}\right), \quad \text{and} \quad \Phi(\alpha z) = \int_{-\infty}^{\alpha z}\mathrm{d}y\phi(y). \quad (1)$$

Here, the shape parameter α controls the degree of asymmetry, with a left (right) skewed density defined by $\alpha < 0 (> 0)$ and the usual normal density by $\alpha = 0$. A location ξ and a scale parameter ω are restored by setting $z = (x - \xi)/\omega$. Model parameters can be estimated, e.g., by the method of moments. The performance of the skewed density (Eq. 1) in fitting daily temperature data was thoroughly assessed in previous works (Simolo et al., 2010, 2011), and is here illustrated in Fig. 2b, c for TX and TN summer anomalies of the AT regional series. Since, as stated above, time averages of second and higher moments are preserved by the evolution of European temperatures, the skewed density (Eq. 1) can be assumed as the fixed-shape PDF underlying daily anomalies. Time dependence can be then imposed on the model PDF through a change in the location parameter ξ only, while keeping the scale and shape parameters constant across the full period and equal to their long-term averages. Hence,

$$z \to z - \frac{\Delta\xi(t)}{\omega}, \quad \Delta\xi(t) = \xi_1 t + \xi_2 t^2, \quad \omega = \bar{\omega} \quad \text{and} \quad \alpha = \bar{\alpha}, \quad (2)$$

where t is the time lapse and a second order term in $\Delta\xi(t)$ accounts for the nonuniform shift of the mean, consistently with observations (see Fig. 2a).

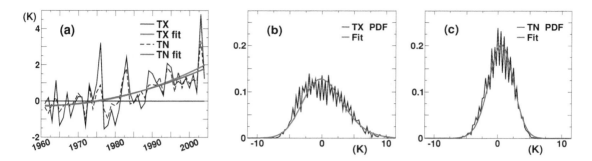

Figure 2. (a) Time evolution of mean summer anomalies for TX (black solid line) and TN (black dashed line) over the AT region, together with their second order polynomial fits (red and blue lines respectively). (**b–c**) Empirical versus model PDFs for detrended summer (**b**) TX anomalies altogether, and similar for (**c**) TN. Model parameters are given in Sect. 4.

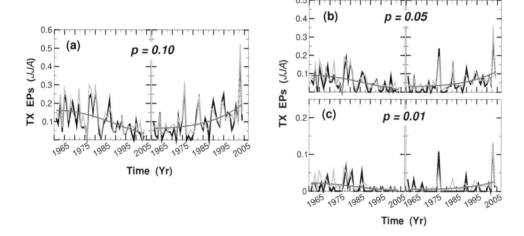

Figure 3. Modeled and observed evolution of TX EPs for AT summer anomalies and probabilities (**a**) $p = 0.10$, (**b**) $p = 0.05$ and (**c**) $p = 0.01$. In all cases, black lines denote the observed frequencies of (left) cold and (right) warm extremes, i.e. TX10 and TX90, and similar. Red lines are expected EPs with location parameter given by Eq. (2), and yellow lines are point-wise, expected EPs (see text).

3.2 Expected EPs

Time-varying frequencies of cold and warm extremes, as measured by the percentile indicators (e.g. TX10 and TX90), have a straightforward, theoretical representation as EPs, respectively

$$F_-(\bar{z}_p, t) = \int_{-\infty}^{\bar{z}_p} dz f(z,t), \qquad \text{and}$$

$$F_+(\bar{z}_{1-p}, t) = \int_{\bar{z}_{1-p}}^{+\infty} dz f(z,t) \qquad (3)$$

for a defined probability p (e.g. $p = 0.10$ for, say, TX10 and TX90). The fixed thresholds, $\bar{z}_p$ and $\bar{z}_{1-p}$ are climatological values, uniquely determined by the condition $F_-(\bar{z}_p, t_0) = F_+(\bar{z}_{1-p}, t_0)$ at a given initial time t_0. In practice, changes over time in the frequency of soft extremes can be explicitly predicted within this framework from observed changes in the distributional properties of daily anomalies, by estimating the appropriate TESD model parameters defined by

Eqs. (1)–(2). The inherent nonlinearity between changes in PDF moments and tails becomes apparent in this formalism, since the time evolution of EPs is controlled by the behavior of the underlying density function. For normal distributed data, for instance, rates of change in EPs scale exponentially in time, even for a uniform shift of the mean, as can be easily proved using Eqs. (1)–(3) with $\alpha = \xi_2 = 0$ (see Simolo et al., 2011 for details).

4 Comparing modeled and observed changes

Warm extremes (i.e. TX > 90th percentile, etc.) have increased at an average rate of several days per decade over all the European areas, and cold extremes (TX < 10th percentile, etc.) correspondingly decreased. Although rates of change in soft extremes are far from being constant over the whole period, linear trends in the AT region amount to, e.g. $+3.1 \pm 0.9$ (-0.3 ± 0.2) summer day-counts per decade in TX90 (TX10), and $+5.2 \pm 0.9$ (-0.5 ± 0.1) in TN90 (TN10).

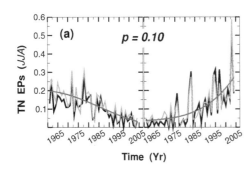

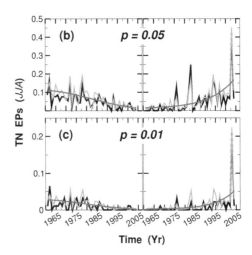

Figure 4. Same as Fig. 3, but for TN EPs.

The frequency of day-counts with JJA anomalies exceeding/not exceeding percentile thresholds are shown as a function of time in Figs. 3–4 (black lines), for both TX and TN in the AT regional series and gradually less moderate extremes.

The expected time-dependent EPs are obtained from Eq. (3) with $p = 0.10, 0.05$, and 0.01, and the same choice of the TESD model parameters defined by Eqs. (1)–(2), derived from observations. That is, the location ξ of the AT summer skewed density is let vary nonlinearly in time with coefficients obtained from fits of TX and TN mean anomalies (see Fig. 2a); estimates of average scale and shape parameters are given by $\bar{\omega}_{TX} = 3.9\,\mathrm{K}$, $\bar{\alpha}_{TX} = 1.3$, $\bar{\omega}_{TN} = 2.4\,\mathrm{K}$, and $\bar{\alpha}_{TN} = -1.1$ (Fig. 2b–c). As seen in Figs. 3–4, theoretical EPs (red lines) nicely fit the observed evolution of both cold and warm extremes, in both TX and TN anomalies, down to $p = 0.01$, thereby outlining the actual rates of change. This, in turn, implies that the observed changes in the frequency of soft extremes are in fact those expected from a forward, nonuniform shift of mean temperatures. This is even more clear if the prescription in Eq. (2) for a continuous time-evolution of the location parameter is replaced with point-wise, annual averages, obtained e.g. by adding back the residuals from the fits of mean anomalies. In all cases the observations are faithfully reproduced by the corresponding point-wise EPs (yellow lines in Figs. 3–4), as witnessed by the large fraction of the variance explained by the model, i.e. $R^2 \sim 0.86, 0.77, 0.49$ (0,88, 0.79, 0.68) for TX (TN) and $p = 0.10, 0.05, 0.01$ respectively, averaging over cold and warm tails. Several unusually intense anomalies in either TX or TN (e.g. the 1976 and 2003 heatwaves) are predicted with values of probabilities very close to the observed ones. This fact again stresses the prominent role of the mean, both long-term trends and fluctuations, in the time-evolution of soft extremes.

Small discrepancies between model predictions and observations can be ascribed to fluctuations in the second and higher moments, not accounted for by the choice of average scale and shape parameters in the TESD model. Nevertheless, these discrepancies are quite small throughout the period, with no pattern, signifying that predictions are not plagued by underestimated trends in variability and skewness.

5 Conclusions

To summarize, we illustrated a simple, theoretical model for predicting changes in the frequency of cold and warm extremes from average distributional properties and changes in daily temperatures. The key-strength of the present approach is its ability to account explicitly for the inherent nonlinearity between PDF moments and tails. A detailed comparison of model predictions and observations over the past decades puts forward the prominent role of the mean in explaining the evolution of extreme summer temperatures over Western Europe, including the occurrence of unusually intense events. In view of this, hypothesis of ongoing changes in the higher moments appear to be neither supported by the data, nor even required.

Acknowledgements. ECA data are archived at the database: http://eca.knmi.nl. This study has been carried out in the framework of the EU project ECLISE (265240).

Edited by: M. Brunet-India
Reviewed by: two anonymous referees

References

Alexander, L. V., Zhang, X., Peterson, T. C., Caesar, J., Gleason, B., Klein Tank, A. M. G., Haylock, M., Collins, D., Trewin, B., Rahimzadeh, F., Tagipour, A., Rupa Kumar, K., Revadekar, J., Griffiths, G., Vincent, L., Stephenson, D. B., Burn, J., Aguilar, E., Brunet, M., Taylor, M., New, M., Zhai, P., Rusticucci, M., and Vazquez-Aguirre, J. L.: Global observed changes in daily climate extremes of temperature and precipitation, J. Geophys. Res., 111, D05109, doi:10.1029/2005JD006290, 2006.

Della-Marta, P. M., Haylock, M. R., Luterbacher, J., and Wanner, H.: Double length of western European summer heat waves since 1880, J. Geophys. Res., 112, D15103, doi:10.1029/2007JD008510, 2007.

Fischer, E. M. and Schär, C.: Future changes in daily summer temperature variability: driving processes and role for temperature extremes, Clim. Dynam., 33, 917–935, doi:10.1007/s00382-008-0473-8, 2009.

Hirschi, M., Seneviratne, S. I., Alexandrov, V., Boberg, F., Boroneant, C., Christensen, O. B., Formayer, H., Orlowsky, B., and Stepanek, P.: Observational evidence for soil-moisture impact on hot extremes in southeastern Europe, Nat. Geosci., 4, 17–21, doi:10.1038/ngeo1032, 2011.

Klein Tank, A. M. G. and Können, G. P.: Trends in indices of daily temperature and precipitation extremes in Europe, J. Climate, 16, 3665–3680, 2003.

Klein Tank, A. M. G., Wijngaard, J. B., Können, G. P., Böhm, R., Demaree, G., Gocheva, A., Mileta, M., Pashiardis, S., Hejkrlik, L., Kern-Hansen, C., Heino, R., Bessemoulin, P., Muller-Westermeier, G., Tzanakou, M., Szalai, S., Palsdottir, T., Fitzgerald, D., Rubin, S., Capaldo, M., Maugeri, M., Leitass, A., Bukantis, A., Aberfeld, R., Van Engelen, A. F. V., Forland, E., Mietus, M., Coelho, F., Mares, C., Razuvaev, V., Nieplova, E., Cegnar, T., Antonio Lopez, J., Dahlstrom, B., Moberg, A., Kirchhofer, W., Ceylan, A., Pachaliuk, O., Alexander, L. V., and Petrovic, P.: Daily dataset of 20th-century surface air temperature and precipitation series for the European Climate Assessment, Int. J. Climatol., 22, 1441–1453, 2002.

Klein Tank, A. M. G., Können, G. P., and Selten, F. M.: Signals of anthropogenic influence on European warming as seen in the trend patterns of daily temperature variance, Int. J. Climatol., 25, 1–16, 2005.

Moberg, A., Jones, P. J., Lister, D., Walther, A., Brunet, M., Jacobeit, J., Alexander, L. V., Della-Marta, P. M., Luterbacher, J., Yiou, P., Chen, D., Klein Tank, A. M. G., Saladie, O., Sigro, J., Aguilar, E., Alexandersson, H., Almarza, C., Auer, I., Barriendos, M., Begert, M., Bergstrom, H., Bohm, R., Butler, C. J., Caesar, J., Drebs, A., Founda, D., Gerstengarbe, F. W., Micela, G., Maugeri, M., Osterle, H., Pandzic, K., Petrakis, M., Srnec, L., Tolasz, R., Tuomenvirta, H., Werner, P. C., Linderholm, H., Philipp, A., Wanner, H., and Xoplaki, E.: Indices for daily temperature and precipitation extremes in Europe analyzed for the period 1901–2000, J. Geophys. Res., 111, D22106, doi:10.1029/2006JD007103, 2006.

Schär, C., Vidale, P. L., Lüthi, D., Frei, C., Häberli, C., Liniger, M. A., and Appenzeller, C.: The role of increasing temperature variability in European summer heatwaves, Nature, 427, 332–336, 2004.

Scherrer, S. C., Appenzeller, C., Liniger, M. A., and Schär, C.: European temperature distribution changes in observations and climate change scenarios, Geophys. Res. Lett., 32, L19705, doi:10.1029/2005GL024108, 2005.

Simolo, C., Brunetti, M., Maugeri, M., Nanni, T., and Speranza, A.: Understanding climate change-induced variations in daily temperature distributions over Italy, J. Geophys. Res., 115, D22110, doi:10.1029/2010JD014088, 2010.

Simolo, C., Brunetti, M., Maugeri, M., and Nanni, T.: Evolution of extreme temperatures in a warming climate, Geophys. Res. Lett., 38, L16701, doi:10.1029/2011GL048437, 2011.

6

Winter-spring cyclonic variability in the Mediterranean-Black Sea region associated with global processes in the ocean-atmosphere system

E. N. Voskresenskaya and V. N. Maslova

Marine Hydrophysical Institute, Sevastopol, Ukraine

Abstract. Using global NCEP/NCAR reanalysis data set on 1000 hPa geopotential height (1948–2006), cyclones in the Mediterranean and Black Sea regions were detected and their main characteristics (frequency, depth, integrated area) were calculated. Analysis of their interannual-multidecadal variability in January-March associated with global processes in the ocean-atmosphere system was done. It was shown that North Atlantic Oscillation (NAO) manifests in the Black Sea region mainly in the variability of frequency of cyclones while in the Mediterranean – in the interannual anomalies of cyclones' depth and area. Joint NAO and El Nino – Southern Oscillation (ENSO) influence is responsible for about 20–45 % of cyclones' frequency variance in the Black Sea region, and in the Mediterranean region for up to 10–25 and 20–30 % of the depth and area variance, accordingly. As a result of using a new approach to study ENSO manifestations based on El Nino classification, correlation coefficients between characteristics of cyclones and Southern Oscillation index (SOI) increase at least twice.

The influence of the Pacific Decadal and Atlantic Multidecadal Oscillations on variability of cyclonic activity in the Mediterranean-Black Sea region manifests in considerable differences of cyclones' characteristics and their typical location.

1 Introduction

The main climatic modes of the global ocean-atmosphere system are the North Atlantic Oscillation (NAO) and El Nino-Southern Oscillation (ENSO) on the interannual scale, while on the decadal-multidecadal scale they are the Pacific Decadal Oscillation (PDO) and Atlantic Multidecadal Oscillation (AMO) as it was shown for example by Barnston and Livezey (1987) and Enfield et al. (2001). Each of these modes has its specific regional manifestations, which are the most pronounced in cold half-year. Cyclonic activity over the Atlantic-European region including the Mediterranean-Black Sea region is an important indicator of regional climate/weather anomalies in this period of year.

NAO manifestations in the interannual variability of frequency of cyclones in the Black Sea region were analyzed in Polonsky et al. (2007). The analysis for different areas of the Mediterranean was done, for example, in Trigo

et al. (2003), Lionello et al. (2006), Guijarro et al. (2006), Pinto et al. (2009). Regional weather-climate anomalies in the Atlantic-European region associated with ENSO were studied, for instance, in Mariotti et al. (2002), Mathieu et al. (2004) whereas ENSO influence on cyclonic activity in the Mediterranean-Black Sea region was not studied yet.

PDO impact on Atlantic-European region was shown in decadal variability of river run-offs and some hydrometeorological characteristics for example by Voskresenskaya (2003), Pekarova et al. (2006). AMO manifestations in the Atlantic-European region were found in large scale temperature and sea level pressure anomalies (Enfield and Mestas-Nunez, 1999), in number and intensity of tropical (Kerr, 2005) and extratropical (Polonsky, 2008) cyclones in the North Atlantic region and some other characteristics.

However, there is still no complex analysis of cyclonic activity in the Mediterranean-Black Sea region in the context of global interannual-multidecadal variability. Taking into account specific seasonality of cyglogenesis in the Mediterranean (as shown in Trigo et al., 1999) and that most prominent manifestations of the above-mentioned signals in the

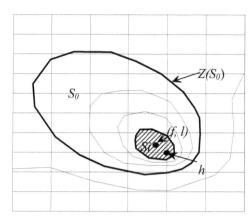

Figure 1. Isobar contour chart. (f, l) – coordinates of the barycenter, $Z(S_0)$ – the last closed isobar contour, S_0 – an area of the figure delimited by the last closed isobar contour, S_i – an area of the figure delimited by the first closed isobar contour, h – a grid point nearest to cyclone's barycenter (after Bardin, 1995).

North Atlantic-European hydrometeorological fields are observed in mature winter-early spring (see for example Polonsky and Voskresenskaya, 1994), the aim of present paper is to study the contribution of NAO, ENSO, PDO and AMO in interannual-multidecadal variability of cyclonic activity in the Black Sea region, Western and Eastern Mediterranean in January–March.

2 Data and methods

On the basis of the daily NCEP/NCAR reanalysis data sets on 1000 hPa geopotential height on a grid $2.5° \times 2.5°$ for the synoptic term 00:00 h GMT in 1948–2006 using the procedure of cyclones' detection by Bardin (1995), time series of the main characteristics (frequency, depth and area) of Mediterranean-Black Sea cyclones were obtained.

The procedure of cyclones' detection is the following. At first the identification algorithm finds all local minima and for each minimum determines a domain of low pressure bounded by closed isobars (see Fig. 1). Then the coordinates of the center of cyclone (f, l) are calculated as a geometrical center of the figure delimited by the first closed isobar contour.

The relative frequency is defined for each considered time interval (specifically one day) as the ratio of the number of cyclones in a given square to the total number detected in the selected period of time (namely one month). The depth of cyclonic eddy is defined as the absolute value of pressure drop between the center of the domain and the last closed isobar (step of isobars is 1 hPa). The area (A) of a cyclonic eddy is defined as the area bounded by the last closed isobar. Obtained cyclones characteristics were averaged over the three regions: the Western Mediterranean, Eastern Mediterranean and Black Sea region (see Fig. 2).

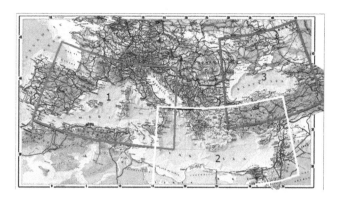

Figure 2. Boundaries of the studied regions: 1 – Western Mediterranean (35–47° N, 6–18° E), 2 – Eastern Mediterranean (29–41° N, 14–38° E), 3 – Black Sea region (37–50° N, 27–45° E).

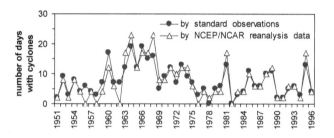

Figure 3. Number of days with cyclonic conditions in the Black Sea region in November obtained using NCEP/NCAR reanalysis data sets and composite-kinematic maps of the Ukrainian Hydrometeorological Service.

Quality control based on comparison of number of days with cyclones selected from NCEP/NCAR reanalysis and the same characteristic from composite-kinematic maps of the Ukrainian Hydrometeorological Service was done. Figure 3 demonstrates a good agreement between graphs of analyzed series (correlation coefficient $r = 0.91$).

As the indicators of global interannual to multidecadal variability, monthly indices of NAO, Southern Oscillation (SOI), PDO and AMO were used.

Standard methods of correlation and composite analyses were used to estimate the connection between variability of the cyclones' characteristics and climatic modes.

As an additional independent source of comparison, maps of the trajectories of cyclones for certain periods were obtained using the Atlas of extratropical storm tracks in 1961–1998 (available online at: http://data.giss.nasa.gov/stormtracks/). Daily NCEP/NCAR reanalysis dataset on the 500 hPa geopotential height were used for composite analyses.

Table 1. Monthly correlation coefficients between the characteristics of cyclones in all studied regions and the NAO index. Coefficients significant on 95 % confidence level are set off in bold font.

Period	Western Mediterranean			Eastern Mediterranean			Black Sea		
	Frequency	Depth	Area	Frequency	Depth	Area	Frequency	Depth	Area
January	−0.08	**−0.34**	**−0.34**	−0.04	**−0.2**	**−0.44**	**−0.53**	**−0.27**	**−0.27**
February	**−0.34**	**−0.22**	**−0.28**	−0.06	**−0.22**	**−0.27**	**−0.3**	**−0.2**	**−0.26**
March	−0.16	**−0.33**	**−0.31**	−0.15	**−0.24**	**−0.37**	**−0.27**	**−0.4**	**−0.46**
Mean	**−0.2**	**−0.3**	**−0.31**	−0.1	**−0.22**	**−0.36**	**−0.38**	**−0.29**	**−0.33**

Table 2. Mean correlation coefficients between the SOI and January–March characteristics of cyclones (lag 5–7 months). All coefficients are significant on 90 % confidence level.

Cyclones' characteristics	Black Sea region	Western Mediterranean	Eastern Mediterranean
Frequency	+0.39	+0.32	+0.34
Area	−0.46	−0.43	−0.25
Depth	−0.33	−0.42	−0.29

3 Interannual variability

Role of the North Atlantic Oscillation and El Nino-Southern Oscillation in the interannual variability of January–March cyclones' characteristics is discussed in this section.

3.1 Manifestations of the NAO and ENSO

The dominant climate pattern of the North Atlantic is NAO. According to the well-known mechanism explained by Hurrell (1995), Polonsky and Voskresenskaya (1994) NAO is responsible for intensification of zonal circulation and for the shift of the North Atlantic storm tracks. In the positive NAO phase storm tracks shift to the Northern Europe and anticyclonic conditions prevail over Central, Southern Europe, and the Mediterranean region, while in its negative phase most of storm tracks displace to the Eastern Europe and Mediterranean region and more anticyclones occur over the Northern Europe.

To estimate the NAO influence on interannual variability of cyclonic activity in the Mediterranean-Black Sea region, a correlation analysis was done. Table 1 demonstrates the negative correlation coefficients between the NAO index and cyclones' characteristics in January–March and confirms the veracity of above-mentioned NAO mechanism of regulation of European climate anomalies. It was corroborated that NAO has significant responses in all studied regions. However, there are some differences in the Black Sea and Mediterranean regions. For the Black Sea region the most significant correlations are between NAO index and frequency of cyclones, while for the Western and Eastern Mediterranean region significant correlations are only for

depth and area of cyclones. There are only few significant correlations with the frequency of cyclones in both parts of the Mediterranean region. This result is corroborated by similar findings by Guijarro et al. (2006). The mean correlation coefficients are in the lowest line of Table 1.

The maximum negative correlation is between frequency of the Black Sea cyclones in January and NAO index ($r = −0.53$). In this case the coefficient of determination (r^2) is 0.28 which means that the NAO explains up to 28 % of the cyclones' frequency variation. Similarly NAO causes up to 15–20 % of the depth and area variability in the Black Sea region in January–March. As for the Mediterranean region NAO is responsible for up to 12 % of depth and area of cyclones' variance in its western part in January and for up to 20 % of area's variance in its eastern part. These results show that NAO manifests stronger in frequency of Black Sea cyclones and depth and area of Mediterranean cyclones. This fact may be explained that Mediterranean is a region of strong cyclogenesis which makes noisy the global climatic mode in comparison with the Black Sea region.

Correlation analysis of SOI and cyclones' characteristics in January-March revealed the maximum significant correlations on the time lag from 5 to 7 months (see Table 2). Maximum of ENSO teleconnection is for the Black Sea and Western Mediterranean regions. Taking into account the magnitudes of assessed coefficients of determination (r^2), we can conclude that from January to March ENSO is responsible for the 15 to 20 % of total variance of cyclones' characteristics after 5 to 7 months. This lag is due to the delay of maximum Pacific sea surface temperature (SST) anomalies after beginning of an ENSO event.

The joint influence of the NAO and ENSO can be estimated as a sum of determination coefficients of each these modes separately on the assumption that they are orthogonal. But we decided to confirm the results of summation of determination coefficients with the analysis of multiple linear regressions by the Eq. (1):

$$C = p_1 + p_2 \cdot I_{NAO} + p_3 \cdot I_{SOI} + \varepsilon, \qquad (1)$$

where C is an outcome variable or one of cyclones' characteristics in this case; p_1 is a constant; p_2 and p_3 are regression coefficients; I_{NAO} and I_{SOI} are independent variables, NAO index and SOI, accordingly; ε is an estimation error.

Calculated multiple coefficients R^2 demonstrate that in January-March the NAO and ENSO are responsible for 20–45 % of the cyclones' frequency variance in the Black Sea region, and in the Mediterranean region for not more than 10–25 and 20–30 % of the depth and area variability, accordingly.

However, the above-mentioned statistics can be improved taking into account the recent results on classification of ENSO events (see for example Horii and Hanawa, 2004; Kao and Yu, 2009; Voskresenskaya and Mikhailova, 2010). This permits to clarify the features of ENSO manifestations in the Mediterranean and Black Sea region.

3.2 Manifestations of different El Nino types

Traditionally Southern Oscillation index (SOI) or oceanic El-Nino index (SST anomalies in region Nino 3.4) are used as the measure of ENSO intensity. The objective classification of El Nino events was done by Voskresenskaya and Mikhailova (2010) using 152-yr time series of SOI and oceanic index. It was shown that there are at least three types of El Nino which differ one from another in the season of beginning, duration and intensity. These are the spring type (SPR), summer-autumn long (SAL) and summer-autumn short types (SAS). Each type has specific manifestations in the North Atlantic – European region. Manifestation in sea level pressure fields are the most important. For instance, "+1" yr of the SPR type is characterized by a positive NAO phase in January, while of the SAL type – by a negative NAO phase and for the SAS type – by neutral NAO.

Unfortunately, in 1948–2008 there were 8 SPR, 6 SAS and 2 SAL ENSO events. That is why in this case we can analyze monthly anomalies of cyclones characteristics only as the case study. The composites of January storm tracks in the North Atlantic are demonstrated in Fig. 4. During SPR type North Atlantic storm tracks shift northeastwardly (which corresponds to the positive NAO phase). SAL events are accompanied by the shift of storm tracks to the Central Europe and Mediterranean region (as in negative NAO phase). During SAS type the North Atlantic storm tracks are slightly shifted to the North-East (as in neutral NAO).

In order to increase enough the number of degrees of freedom, special time series of January–March cyclones' charac-

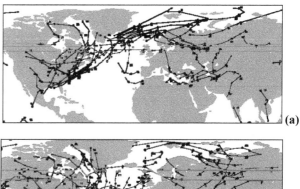
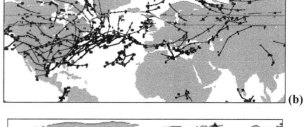
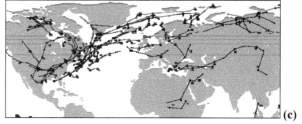

Figure 4. Typical January storm tracks for the "+1" year of the each type of El Nino on the example of the pair of events: **(a)** spring type – 1977, 1995; **(b)** summer-autumn long type – 1969, 1987; **(c)** summer-autumn short type – 1983, 1998.

teristics were obtained in the following way. Monthly characteristics from January to March were taken consequently for all years with certain El Nino type. Taking into account time lag of 5–7 months, the analogous series of SOI were obtained: Aug–Oct (lag = 5 months), Jul–Sep (lag = 6 months), Jun–Aug (lag = 7 months). This procedure permits to increase enough the number of degrees of freedom of the time series of two types of El Nino (Spring and Summer–Autumn Short) within the correlation analysis. As a result, correlation coefficients between SOI and cyclones' characteristics increased to +0.7 to +0.8 for the SPR type of El Nino and to −0.8 to −0.9 for the SAS type. SAL type was not analyzed because of the mentioned reason. The strong increase of correlation coefficients can be explained as following. If we consider the total number of ENSO events, in this case manifestations of one El Nino type smooth manifestations of the other.

Obtained results highlight that a correct study of ENSO manifestations is possible only taking into account the differences of ENSO types.

4 Decadal variability

Further analysis was focused on decadal and multidecadal variability of cyclones' characteristics associated with the AMO and PDO. Only one period of these oscillations was registered in 1948–2006. That is why composite analysis for their negative and positive phases was done.

4.1 PDO manifestations

As it was shown for example in Mantua et al. (1997), PDO is the most prominent quasi-periodical (about 20 to 30 yr) global climate mode located between 20 and 60° N in the Pacific Ocean. It is characterized by alternation of large-scale negative and positive SST anomalies. To estimate PDO manifestations in cyclonic activity in the studied region, composite values of monthly cyclones' characteristics in the periods of negative (1948–1976) and positive (1977–2001) PDO phases were calculated. As an example cyclones' frequency is shown in Fig. 5. In certain winter-spring months cyclonic activity in all parts of the Mediterranean-Black Sea region is significantly (relative to the interannual variability) higher during the negative PDO phase than during the positive one. Particularly, in the Black Sea region January–March frequency of cyclones was twice higher during the negative PDO phase. This result is significant on the 95 % confidence level. In negative PDO phase December–April composite values of the area of the Black Sea cyclones exceeded those in positive phase by absolute values, significantly only in February by ~500 × 10³ km² (90 % confidence level). In the negative PDO phase the depth of cyclones was ~5 hPa greater than in the positive phase (on the 90 % confidence level in February). For the Mediterranean cyclones the response is lower, but still evident in certain months. The frequency of cyclones during the negative PDO phase in the Western Mediterranean was up to ~20 % higher in December and March and in the Eastern Mediterranean – in March and October (on 80 % confidence level). Composite values of the other cyclones' characteristics (depth and area) in both parts of the Mediterranean region were greater by absolute values also in winter-spring months during the negative PDO phase.

4.2 AMO manifestations

Climatic signal AMO (SST anomaly over the North Atlantic) is the analogue of multidecadal NAO mode. It is characterized by a shift of the North Atlantic storm tracks similar to NAO but with the opposite sign (see for instance Polonsky, 2008). During the positive NAO phase the Icelandic Low and Azores High are mostly shifted to the Northeast, while during the positive AMO phase to Southwest. Associated trajectories of North Atlantic cyclones shift to the Northern Europe or to the Southern Europe and Mediterranean region.

In the result of composite analyses we confirm this abovementioned scheme using NCEP/NCAR reanalysis dataset,

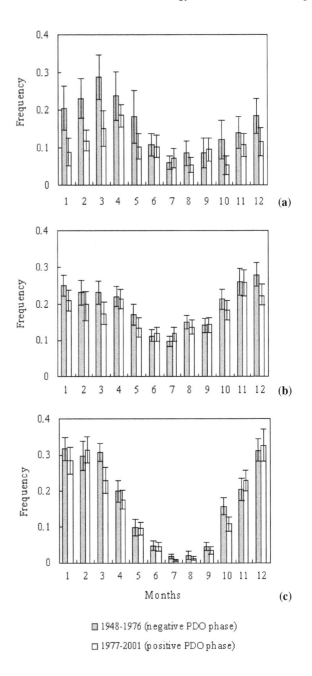

Figure 5. Composite values of the frequency of cyclones in the (a) Black Sea region, (b) Western Mediterranean and (c) Eastern Mediterranean during the negative (1948–1976) and positive (1977–2001) PDO phase. Confidence level corresponds to the 80 % of statistical significance.

namely composite anomalies of the 500 hPa geopotential height field and the storm tracks in the negative (1971–1975) and positive (1994–1998) AMO phases. During the negative AMO phase there are significant negative anomalies of the 500 hPa geopotential height in the vicinity of the Icelandic Low and positive anomalies in the Azores High (not shown). For this situation the NAO index is positive. That is why

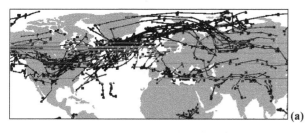

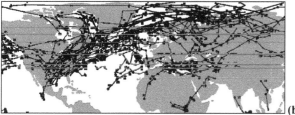

Figure 6. January storm tracks during the periods of the (**a**) minimum (1971–1975) and (**b**) maximum (1994–1998) AMO index.

North Atlantic storm tracks move to the North-Eastern part of the North Atlantic, as shown in Fig. 6a. In the Mediterranean region and Central Europe cyclonic activity is low. During the positive AMO phase Icelandic Low is characterized by the positive anomalies of the 500 hPa geopotential height and there are negative anomalies in the Azores High vicinity. Therefore, centers of action of atmosphere in the North Atlantic are weak and the NAO is negative. Figure 6b illustrates that during this period cyclonic activity is intensive in the Mediterranean-Black Sea region.

The described mechanism is also confirmed by the positive sign of correlation coefficients between the characteristics of Mediterranean-Black Sea cyclones and unsmoothed AMO index. Correlation analysis was done separately for the periods of the positive (1948–1960) and negative (1966–1995) AMO phases and for the periods of transition from one phase to another (1960–1975, 1990–2006, accordingly). The maximum correlation coefficients significant on 95 % confidence level (relative to the interannual variability) reach 0.3–0.6 for all analyzed time intervals. On the basis of coefficient of determination it is possible to conclude that up to 36 % of variance of the cyclones' characteristics is associated with the AMO influence.

5 Conclusions

As a result of analyses of interannual-multidecadal January–March cyclonic variability in the Mediterranean-Black Sea region associated with NAO, ENSO, PDO and AMO the following results were formulated.

NAO manifests mainly in cyclones' frequency variability in the Black Sea region and in depth and area variability in the Western and Eastern Mediterranean. The joint NAO and ENSO influence is responsible for 20–45 % of cyclones' frequency variance in the Black Sea region, while in the Mediterranean region for up to 10–25 and 20–30 % of the depth and area variance, accordingly.

Manifestations of three classified El Nino types in cyclones' variability cardinally differ one from another. It is highlighted that ENSO classification is essential in the correct analysis of regional ENSO manifestations.

PDO and AMO have significant (relative to the interannual variability) manifestations in the decadal variability of cyclonic activity in the Mediterranean-Black Sea region from December to March. The negative PDO phase is accompanied by the intensification of cyclonic activity in the comparison with the positive PDO phase in the studied regions. Particularly, cyclones' frequency in the Black Sea region double in the negative PDO phase from January to March. Our research confirms that the AMO considerably influences on the behavior of storm tracks in the Mediterranean-Black Sea region in winter-spring months (December–March), with correlation coefficients up to +0.3–+0.6.

Acknowledgements. The authors thank referees and editor for evaluating this paper.

Partly results were obtained within the grant FP7 Collaborative Project "THESEUS", Grant Agreement No. 244104.

Edited by: S. Gualdi
Reviewed by: two anonymous referees

References

Bardin, M. Yu.: Variability of cyclonic characteristics in the middle troposphere of the North hemisphere midlatitudes, Meteorology and hydrology, 11, 24–37, 1995 (in Russian with English abstract).

Barnston, A. and Livezey, R.: Classification, seasonality and persistence of low-frequency atmospheric circulation patterns, Mon. Weather Rev., 115, 6, 1083–1126, 1987.

Enfield, D. and Mestas-Nunez, A. M.: Multiscale variability in global SST and their relationships with tropospheric climate patterns, J. Climate, 9, 2719–2733, 1999.

Enfield, D. B., Mestas-Nunez, A. M., and Trimble P. J.: The Atlantic Multidecadal Oscillation and its relationship to rainfall and river flows in the continental U.S., Geophys. Res. Lett., 28, 2077–2080, 2001.

Guijarro, J. A., Jansà, A., and Campins, J.: Time variability of cyclonic geostrophic circulation in the Mediterranean, Adv. Geosci., 7, 45–49, doi:10.5194/adgeo-7-45-2006, 2006.

Horii, T. and Hanawa, K.: A relationship between timing of El Nino onset and subsequent evolution, Geophys. Res. Lett., 31, 1634–1648, 2004.

Hurrell, J. W.: Decadal trends in the North Atlantic Oscillation and relationships to regional temperature and precipitation, Science, 269, 676–679, 1995.

Kao, H. Y. and Yu, J. Y.: Contrasting eastern Pacific and central Pacific types of ENSO, J. Climate, 22, 615–632, 2009.

Kerr R. A.: Atlantic climate pacemaker for millennia past, decades hence? Sience, 5731, 41–42, 2005.

Lionello, P., Bhend, J., Buzzi, A., Della-Marta, P. M., Krichak, S., Jansà, A., Maheras, P., Sanna, A., Trigo, I. F., and Trigo, R.: Cyclones in the Mediterranean region: climatology and effects on the environment, in: Mediterranean Climate Variability, edited by: Lionello, P., Malanotte-Rizzoli, P., and Boscolo, R., Elsevier, Amsterdam, 324–272, 2006.

Mantua, N. J., Hare S. R., Zhang Y., Wallace J. M., and Francis R. C.: A Pacific interdecadal climate oscillation with impacts on salmon production, BAMS, 78, 1069–1079, 1997.

Mariotti, A., Zeng, N., and Lau, K.-M.: Euro-Mediterranean rainfall and ENSO – a seasonally varying relationship, Geophys. Res. Lett., 29, 1621, doi:10.1029/2001GL014248, 2002.

Mathieu, P. P., Sutton, R. T., Dong, B., and Collins, M.: Predictability of winter climate over North Atlantic European region during ENSO events, J. Climate, 17, 1953–1974, 2004.

Pekarova, P., Miklanek, P., and Pekar, J.: Long-term trends and runoff fluctuations of European rivers, Fifth FRIEND World Conference (Havana, Cuba, November 2006), IAHS Publ., p. 308, 2006.

Pinto, J. G., Zacharias, S., Fink, A. H., Leckebusch, G. C., and Ulbrich, U.: Factors contributing to the development of extreme North Atlantic cyclones and their relationship with the NAO, Clim. Dynam., 32, 711–737, 2009.

Polonsky, A. B.: Atlantic multidecadal oscillation and its manifestations in the Atlantic-European region, Marine Hydrophysical Journal, 4, 47–58, 2008 (in Russian with English abstract).

Polonsky, A. and Voskresenskaya, E.: Interannual variability in the Tropical and North Atlantic associated with the Pacific Ocean ENSO events, Contemporary Climatology, Proc. of the Meeting of the Commission on Climatology of the IGU, Brno, 467–472, 1994.

Polonsky, A. B., Bardin, M. Yu., and Voskresenskaya, E. N.: Statistical characteristics of cyclones and anticyclones over the Black Sea in the second half of the 20th century, Physical Oceanography, 17, 348–359, 2007.

Trigo, I. F., Davies, T. D., and Bigg, G. R.: Objective climatology of cyclones in the Mediterranean region, J. Climate, 12, 1685–1696, 1999.

Trigo, R. M., Pozo-Vazquez, D., Castro-Diez, Y., Osborn, T., Gamis-Fortis, S., and Esteban-Parra, M. J.: NAO influence on precipitation, river flow regimes and hydroelectric power generation in the Iberian peninsula, Geophys. Res. Abstr., 5, 05494, 2003.

Voskresenskaya, E. N.: Variability of the European rivers' runoff associated with global climate processes, Monitoring systems of environment, CSW of NASU, MHI, 144–147, 2003 (in Russian with English abstract).

Voskresenskaya, E. N. and Mikhailova, N. V.: El-Niño classification and climate/weather anomalies in the Black Sea region, Monitoring systems of environment, Reports of the NAS of Ukraine, 3, 120–124, 2010 (in Russian with English abstract).

7

Analysis of climate hazards in relation to urban designing in Iran

S. Alijani[1] **and B. Alijani**[2]

[1]Department of Urban Designing, Islamic Azad University, Tehran, Iran
[2]Center of Excellence for Spatial Analysis of Environmental Hazards, Tarbiat Moallem University, Tehran, Iran

Abstract. In order to study the climate hazards, daily rainfall and temperature data of 61 weather stations over the country were obtained from the Meteorological Organization of Iran for the 1951–2007 period. The following indices are defined as indications of climate hazards: sultriness of the air or the heat index, cold days with minimum temperature below −5 °C, warm days with maximum temperature above 32 °C, the share of extreme rain days from the annual rainfall. The annual frequencies of these indices are analyzed and the overall hazard index is computed using the Analytical Hierarchical Process method.

The results show that the southern coastal areas and central deserts are the most hazardous parts of the country, whereas, the northern Caspian coastal lands and mountainous regions experience lower hazard alerts. The problem of the northern parts is the cold days and that of the southern areas is the hot and humid days. Despite the relatively equal occurrence of torrential rains over the country, they are more harmful in the south than in the other parts of the country.

1 Introduction

Phenomena like flash floods or heat waves and their possible erratic occurrence are important climate hazards (Alijani et al., 2008) for designing and planning the urban centers, predominantly in arid regions of the globe. The impact of these climate hazards is serious and destructive in the dense and fragile populated settlements. They usually damage the urban infrastructures; create polluted environment, and unhealthy living conditions. The degree of damage (climate risk) of these hazards depends among other factors such as vulnerability and exposure of the environment, on their extent, severity, and especially their probability of occurrence (Gwilliam et al., 2006).

The concern about the impact of the physical environment on the urban development can be traced more than 2000 years back to the time of Vitruvius who has discussed in his book "*The Ten Books on Architecture*" the importance of climate on the layout of cities (Morgan, 1960). Since then many publications have focused on the climate impact of the city. Grunthal and his colleagues (2006) have analyzed the damage of three main hazards in the city of Cologne, Germany and present the suitable methods to analyze their impact. The works of Olgyay (1963) and Givoni (1998) have presented good tools for measuring human comfort indices in the urban areas. Scientists of some institutions such as the Meteorological Institute of the University of Freiburg, Germany have done remarkable researches in the area of climate and human health. Works of A. Matzarakis (such as Matzarakis and Mayer, 2000) are good examples of this kind. For more information about the interaction between city and the environment the readers are referred to works of Oke (e.g., Oke, 1987).

During the past one hundred years the global temperature has increased about 0.74 °C (IPCC, 2007; Zhang et al., 2005). In Iran, the outcome of this global warming is the increasing frequency of extreme events such as cold spells, heat waves, torrential rains, or prolonged intensive droughts (Rahimzadeh et al., 2009); hence increasing danger and harm to the built environments and people. The analysis of intensity and frequency of these extreme events is a very important step in the process of planning and designing the urban areas especially in the semi arid climate of Iran. In these areas climate is very fragile and a sudden change may cause destructive outcomes. During the 2000–2008 period droughts and floods were the first in terms of economic loss and number of affected people in Iran (http://www.emdat.be/search-details-disaster-list).

Table 1. Indices used to identify extreme climatic events.

Heat Index (HI)	$ET = -42.379 + 2.04901523T + 10.14333127R - 0.22475541TR - 6.83783 \times 10^3 \, T^2 - 5.481717 \times 10^2 \, R^2 + 1.22874 \times 10^3 \, T^2R + 8.5282 \times 10^4 \, TR^2 - 1.99 \times 10^6 \, T^2R^2$
Cold Days (CD)	Days with minimum temperature below $-5\,°C$
Warm Days (WD)	Days with maximum temperature above $32\,°C$
Extreme Rains Percentage (ERP)	Ratio of extreme rains from the annual rain

The spatial and temporal variability is the intrinsic characteristic of the climate of Iran. According to Tagavi and Mohammadi (2007) during the past years, the frequency of cold events has decreased while that of warm events has increased. Kary (2010) has confirmed the significant change of the heat waves and cold spells in Tehran during past years. All these studies and others such as the work of Rasooli (2004) about the spatial analysis of cold winds over the northwestern Iran and that of Kaviani and his colleagues (2004) about the effective temperature over the country show the variability and uncertainty of the climate over the country. Therefore it is very important for the planners, designers, and managers of the urban environments to pay attention to the climatic considerations. In any part of the country its especial climate should be understood and the extreme cases ought to be analyzed. But up to now no comprehensive work has been carried out in this field; the task which this research has undertaken. The following section is devoted to the description of data and methodology. The main outcome of this investigation is presented in the Results section. The paper is closed with a brief conclusion.

2   Data and methodology

A great deal of literature has been carried out about the definition and impacts of the climate hazards on the environment and people. The thresholds are defined differently. Some researchers use the percentile thresholds such as the above 90th percentile or the below 5th percentile (Hayhoe et al., 2004; Díaz et al., 2006). While the others may use absolute values such as 32 °C or daily rains above 30 mm. Most researchers use both methods (Bell et al., 2004; Frich et al., 2002; Haylock and Nicholis, 2000; Bärring and Persson, 2006). Considering all these different methods the following four indices (Table 1) are defined in this study.

The first index in Table 1 is a combination of temperature (T, in degrees Fahrenheit) and relative humidity (RH, in decimal format) called Heat Index (HI). It is defined as the overall effect of the combination of temperature and relative humidity (Steadman, 1979). For, most of the time the atmosphere is calm in the urban environment and the amount of the relative humidity controls the effect of the temperature. The higher relative humidity increases the adverse effect of the temperature (Zaboli, 2010). We use this index to mea-

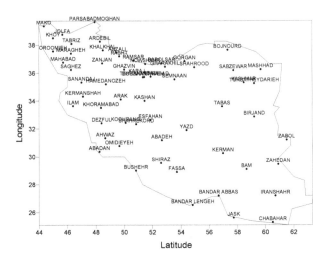

Figure 1. The location of the study stations.

sure the heat stress on humans. Days with HI value above 80 are considered the days with heat stress causing adverse impacts such as sunstroke, muscle cramps and other related problems. These values are averaged on the annual basis for each station. The days with a minimum temperature below −5 °C are defined as days with cold stress, because in the dry environment of Iran the temperatures above −5 °C are more common and the environment is adopted to it. In the same way the warm days are defined as days with maximum temperature above 32 °C. The ratio of the rainfall of extreme daily rains from the annual rainfall is defined as the Extreme Rains Percentage (ERP). A day with extreme rainfall is defined as a day with rainfall above the 90th percentile of the station daily rainfall time series during the study period; dry days are not included in this time series. The defined indices are extracted from the daily data of 61 synoptic stations during the period starting from 1951 to 1985 and ending to 2007. The studied stations were shown in Fig. 1.

The overall hazardiness of each station is calculated according to the Analytical Hierarchical Processing (AHP) method, developed by Saati (Saati, 1980). In this method each of the four indices is weighted according to its importance. The HI is considered as the most important index for humans. The ERP index is defined as the second important index in the urban planning and designing, because, the extreme rains usually far exceed the capacity of cities

Table 2. Weighting of indices.

	HI	ERP	CD	WD	Weights	Normalized Weights
HI	1	3	5	7	3.20	0.56
ERP	0.33	1	3	5	1.49	0.26
CD	0.2	0.33	1	3	0.67	0.12
WD	0.1429	0.2	0.33	1	0.31	0.05
					5.67	1

Table 3. Weighting of four classes of Heat Index.

	201–262	139–200	76–138	13–75	Weights	Normalized Weights
201–262	1	3	5	7	3.20	0.56
139–200	0.33	1	3	5	1.49	0.26
76–138	0.2	0.33	1	3	0.67	0.12
13–75	0.1429	0.2	0.33	1	0.31	0.05
					5.67	1

Table 4. The final scores used for classifying indices.

First Level	Score	Second Level	Score	Final Score
HI	0.56387	201–262	0.564428	0.318
HI	0.56387	139–200	0.263004	0.148
HI	0.56387	76–138	0.117619	0.066
HI	0.56387	13–75	0.054948	0.031
ERP	0.263004	0.494–0.549	0.564428	0.148
ERP	0.263004	0.438–0.493	0.263004	0.069
ERP	0.263004	0.382–0.437	0.117619	0.031
ERP	0.263004	0.326–0.381	0.054948	0.014
CD	0.117619	84–109	0.564428	0.066
CD	0.117619	56–83	0.263004	0.031
CD	0.117619	28–55	0.117619	0.014
CD	0.117619	0–27	0.054948	0.006
WD	0.054948	309–411	0.564428	0.031
WD	0.054948	206–308	0.263004	0.014
WD	0.054948	103–205	0.117619	0.006
WD	0.054948	1–102	0.054948	0.003

and cause destructive floods and damage buildings and other infrastructures. Their destructive effect is manifold in dry environments such as Iran due to their rare and erratic occurrence. The frequency of cold spells comes as the third important index in urban planning and designing. Colder and biter temperatures are very harmful for people and infrastructures, especially in the arid climate of Iran. The Warm Days are scored as the least important of all, because, high temperatures are very frequent in arid areas and people and structures are adopted to them. The pair wise preferences of e.g. the indices are defined from 7 for HI to 1 for WD (Saati, 1980) and shown in Table 2. According to this table, the preference of HI is seven times WD, the preference of ERP is 3 times CD and 5 times WD. In each row the preference scores are geometrically averaged and written in the Weights column. The normalized weights are shown in the final column. This column indicates that the importance of HI in causing hazard for the environment is 5.5 times CD (0.56/0.12 = 5.5). Each index is divided into four subgroups and processed according to the same method described in Table 2; subgroups of HI are shown as an example in Table 3.

Each index is divided into four classes and weighted according to their frequency during the study period (Table 4).

The final preference scores are computed by multiplying the normalized weights of the indices and their subgroups (Table 4). The scores of the final column of Table 4 are given to the related value of each index in the actual data table as exemplified in Table 5.

The hazard score of each station (last column of Table 5) is added up from the weights given to the values of the indices in that station according to Table 4. The spatial distribution of hazard indices is interpolated and drawn with the Inverse Distance Weight method.

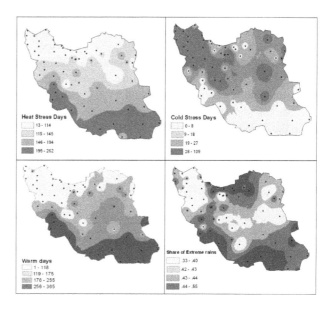

Figure 2. Spatial distribution of hazard indices.

3   Results

The spatial distributions of the selected indices are shown in Fig. 2. The spatial distributions of all maps are divided into four quintiles. According to the HI map, the frequency of these days is high in the south and decreases toward the north. The southern parts of the country suffer from sultry weather. The sultry conditions develop harm to human body and cause susceptible environment for different diseases. Therefore it is a severe limitation for the urban development and expansion. The same spatial pattern exists in the warm days map. It is obvious from these two maps that the southern quarter of the country needs more energy to mitigate this hot and humid environment.

Table 5. An example of weighting of each station.

Station	Values				Weights				Hazard Score
	HI	CD	WD	ERP	Hid	ERP	CD	WD	
ABADAN	224	0	360	0.46	0.32	0.069	0.006	0.031	0.426
ABADEH	126	26	75	0.41	0.07	0.031	0.006	0.003	0.11
ABALI	13	109	1	0.45	0.03	0.069	0.066	0.003	0.168
AHWAZ	221	0	361	0.41	0.32	0.031	0.006	0.031	0.388

The spatial distribution of cold days is in contrast with that of the warm or sultry days. On this map the north-western parts and the central Zagros are suffering from cold weather. Here also, more energy is required to compensate this cold environment. According to these maps of warm and cold days both the southern and northern parts of the country need more energy to provide a suitable environment for urban dwellers throughout the year. The share of the torrential rains from the annual rainfall varies from 33 to 55 % over the country. Thus all of the country is vulnerable for torrential rains with some emphasize to the coastal lands. Urban floods are severe and frequent in the coastal areas, especially in the very dry and erratic climate of the South.

The comparison of these four maps indicates that at any place in the country there exists at least one climatic problem for urban activities. But places such as the southern coastal lands experience multiple hazards of hot and sultry climate together with torrential rains. These climate hazards are seasonal in most of the country, but dominate all year around in the southern coastal areas.

4 Hazard map

The computed susceptibility or vulnerability to the climate hazards of each station as exemplified in Table 5, are given in Table 6 and drawn in Fig. 3. According to this figure the most vulnerable places of the country are the southern coastal areas. Here most of the year the hot and humid climate is dominant. The daily rainfall is very erratic and torrential. The region suffers most of the year from either torrential rains or very serious droughts. The colder areas of the north and mountains and drier central parts of the country do not exhibit a harsh environment. The vulnerability of these parts is about one third of the southern regions (0.18/0.56 = 0.33). The comparison of the ERP map of Fig. 2 with the hazard map of Fig. 3 shows that the relatively higher percentage of torrential rains in the southern coasts of the Caspian Sea does not lead to the hazardous conditions, because, these heavy rains are more common in the Caspian coast and the region is adopted to them. But due to their erratic and infrequent occurrence, the torrential rains always cause serious problems in the South.

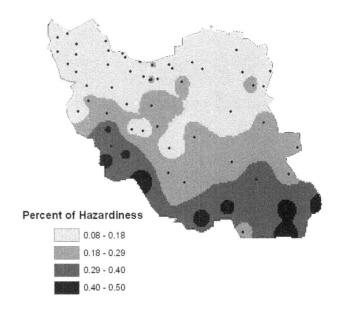

Percent of Hazardiness

☐ 0.08 - 0.18
▨ 0.18 - 0.29
▦ 0.29 - 0.40
■ 0.40 - 0.50

Figure 3. Spatial variation of climate hazard in Iran.

An important point to be mentioned is that our measure of climatic vulnerability is based on the annual frequency of the extreme climatic conditions. Because, the persistence of hazardous conditions cause more problem than their intensity. The prolonged cold weather of winter 2006 caused very high damage over the country or the persisting snow storm of Gilan Province in winter 2005 damaged most of the urban structures of the area. As a result, the southern regions with frequent extreme climate conditions show up as hazardous parts of the country; the fact that exists in reality. In the northern parts of the country, the severe cold conditions or harmful sultry conditions do not persist for long periods. In these areas, due to the dominance of westerlies during the cold season, the hazardous cold conditions last short times of maybe 5 to 7 days (Alijani, 2002), after which a sunny weather warms the environment and lowers the degree of vulnerability.

The hazard map indicates that there is climatic vulnerability for urban centers all over the country. The urban managers and planners or designers should take into consideration the cold weather problems such as street frosts and snow accumulation, winter season energy consumption in the

Table 6. The climatic hazard score of the studied stations.

Station	Hazard Score	Station	Hazard Score	Station	Hazard Score
ABADAN	0.426	GHAZVIN	0.151	NOWSHAHR	0.187
ABADEH	0.11	GORGAN	0.111	OMIDIEYEH	0.426
ABALI	0.168	HAMEDANOZEH	0.17	OROOMIEH	0.173
AHWAZ	0.388	ILAM	0.113	PARSABADMOGHAN	0.096
ANZALI	0.108	IRANSHAHR	0.426	RAMSAR	0.187
ARAK	0.156	JASK	0.256	RASHT	0.11
ARDEBIL	0.13	JOLFA	0.121	SABZEWAR	0.176
BABOLSAR	0.148	KARAJ	0.11	SAGHEZ	0.118
BAM	0.371	KASHAN	0.239	SANANDAJ	0.135
BANDAR ABBAS	0.426	KASHMAR	0.231	SEMNAAN	0.201
BANDAR LENGEH	0.426	KERMAN	0.201	SHAHREKORD	0.135
BIRJAND	0.193	KERMANSHAH	0.201	SHAHROOD	0.148
BOJNOURD	0.118	KHALKHAL	0.13	SHIRAZ	0.193
BUSHEHR	0.505	KHORAMABAD	0.193	TABAS	0.089
CHABAHAR	0.426	KHOY	0.135	TABRIZ	0.135
DEZFUL	0.426	KOOHRANG	0.13	TEHRANMEHRABAD	0.184
DOUSHANTAPEH	0.193	MAHABAD	0.118	TORBATHEYDARIEH	0.078
ESFAHAN	0.193	MAKO	0.13	YAZD	0.201
FASSA	0.193	MARAGHEH	0.118	ZABOL	0.239
GHARAKHIL	0.151	MASHHAD	0.118	ZAHEDAN	0.231
				ZANJAN	0.118

north. In the south and central parts, the streets and buildings should be designed in accordance with hot and sultry conditions. The northern cities require sheltered and narrow streets to protect the pedestrians from cold weather. But in the south the streets should be wide enough to cope with prolonged hot summers. The buildings should be designed in a way to block the penetration of excessive solar radiation inside. In the north the buildings should allow the solar radiation inside via large windows and solar oriented facades. The designing process should proceed in a way to prevent fossil energy use and get more benefit from renewable solar and wind energy. In this way air pollution will decrease and global warming may cease. Most of the cities require a high capacity drainage mechanism such as underground drainage systems to prevent the sudden torrential rains and floods.

5 Conclusions

We studied the spatial distribution of climate hazards over Iran in relation to the planning and designing of sustainable urban environments; The results show that the climate hazards and vulnerability are present all over the country but with different degrees. The degree of climate risk decreases from south toward the northern parts. The most hazardous part of the country is the southern part, where the hot and humid summers cause harsh living conditions. In the northern parts the main climate hazard is the cold and icy conditions of the winter season. The extreme rains are harmful in the

south but not in the north. The spatial variability of the climate hazards indicates that different urban structures should be planned for the different parts of the country. In order to plan and manage a sustainable city, the planners and designers should provide summer comfort in the south and winter comfort in the northern parts of the country.

Edited by: M. Dolinar
Reviewed by: M. Farajzadeh and another anonymous referee

References

Alijani, B.: Variations of 500 hPa flow patterns over Iran and surrounding areas and their relationship with the climate of Iran, Theor. Appl. Climatol., 72, 41–54, 2002.

Alijani, B., O'Brien, J., and Yarnal, B.: Spatial analysis of precipitation intensity and concentration in Iran, Theor. Appl. Climatol., 94, 107–124, 2008.

Bärring, L. and Persson, G.: Influence of climate change on natural hazards in Europe. Natural and technological hazards andrisks affecting the spatial development of European regions, Geol. S Finl., 42, 93–107, 2006.

Bell, J., Sloan, L. C., and Snyder, M. A.: Regional Changes in Extreme Climatic Events: A Future Climate Scenario, J. Climate, 17, 81–87, 2004.

Díaz, J., García-Herrera, R., Trigo, R. M., Linares, C., Valente, M. A., De Miguel, J. M., Frich, E., Alexander, P. L. V., Della-Marta, P., Gleason, B., Haylock, M., Klein Tank, A. M. G., and Peterson, T.: Observed coherent changes in climatic extremes during the second half of the twentieth century, Clim. Res., 19, 193–212, 2006.

Frich, P., Alexander, L. V., Della-Marta, P., Gleason, B., Haylock, M., Klein Tank, A. M. G., and Peterson, T.: Observed coherent changes in climatic extremes during the second half of the twentieth century, Clim. Res., 19, 193–212, 2002.

Givoni, B.: Climate Considerations in Building and Urban Design, Van Nostrand Reinhold, New York, 1998.

Grunthal, G., Thieken, A. H., Schwarz, J., Radtke, K. S., Smolka, A., and Merz, B.: Comparative Risk Assessments for the City of Cologne – Storms, Floods, Earthquakes, Nat. Hazards, 38, 21–44, doi:10.1007/s11069-005-8598-0, 2006.

Gwilliam, J., Fedeski, M., Lindley, S., Theuray, N., and Handley, J.: Methods for assessing risk from climate hazards in urban areas. Proceedings of the Institution of Civil Engineers, Municipal Engineer, 159, 245–255, 2006.

Hayhoe, K., Kalkstein, L., Moser, S., and Miller, N.: Rising Heat and Risks to Human Health: Technical Appendix, Union of Concerned Scientists Publications,Cambridge, MA, USA, 2004.

Haylock, M. and Nicholis, N.: Trends in extreme rainfall indices for an updated high quality data set for Australia, 1910–1998, Int. J. Climatol., 20, 1533–1541, 2000.

IPCC: Assessment Report 4, Climate Change 2007: Synthesis Report, 2007.

Kary, S.: Spatial and Temporal analysis of heat waves and cold spells in Tehran urban area, M.S. Thesis, Tarbiat Moalem University, Tehran, Iran, 2010.

Kaviani, M. R., Gayour, H. A., and Payandeh, N.: Evaluating the effective temperature over Iran, Quarterly Journal of Geographical Researches, 80, 13540–13549, 2004.

Matzarakis, A. and Mayer, H.: Atmospheric conditions and human thermal comfort in urban areas, Proceedings of 11th Seminar on Environmental Protection "Environment and Health", 20–23 November 2000, Thessaloniki, Greece, 155–166, 2000.

Morgan, M. H.: Vitruvius, The Ten Books On Architecture, Dover Publications, New York, 1960.

Olgyay, V.: Design with Climate, Princeton University Press, New Jersey, 1963.

Oke, T. R.: Boundary layer climates, 2nd Edn., Routledge, New York, 1987.

Rahimzadeh, F., Asgari, A., and Fattahi, E.: Variability of extreme temperature and precipitation in Iran during recent decades, Int. J. Climatol., 29, 329–343, 2009.

Rasooli, A. A.: Spatial modeling the wind cooling in the northwest of Iran, Quarterly Journal of Geographical Researches, 80, 13559–13579, 2004.

Saati, T.: The Analytic Hierarchy Process: Planning, Priority Setting, Resource Allocation, ISBN 0-07-054371-2, McGraw-Hill, New York, 1980.

Steadman, R. G.: The assessment of sultriness. Part I: A temperature-humidity index based on human physiology and clothing science, J. Appl. Meteorol., 18, 861–873, 1979.

Tagavi, F. and Mohammadi, H.: Studying the return period of climatic extreme events in order to understand their environmental effects, J. Environ. Stud., 43, 11–20, 2007.

Zaboli, S.: Climatic potentials of kerman province of Iran for tourism activities, M.A. Thesis, Tabarestan University, Chalus, Iran, 2010.

Zhang, X., Aguilar, E., Sensoy, S., Melkonyan, H., Tagiyeva, U., Ahmed, N., Kutaladze, N., Rahimzadeh, F., Taghipour, A., Hantosh, T. H., Albert, P., Semawi, M., Karam Ali, M., Halal, M., Al-Shabibi, S., Al-Oulan, Z., Zatari, T., Al Dean Khelet, I., Hamoud, S., Sagir, R., Demircan, M., Eken, M., Adiguzel, M., Alexander, L., Peterson, T. C., and Wallis, T.: Trends in the Middle East climate extreme indices from 1950 to 2003, J. Geophys. Res., 110, D22104, doi:10.1029/2005JD006181, 2005.

8

Homogenising time series: beliefs, dogmas and facts

P. Domonkos

Centre for Climate Change (C3), Geography Dept., University Rovira i Virgili, Campus Terres de l'Ebre, C. Betánia 5, Tortosa, 43500, Spain

Abstract. In the recent decades various homogenisation methods have been developed, but the real effects of their application on time series are still not known sufficiently. The ongoing COST action HOME (COST ES0601) is devoted to reveal the real impacts of homogenisation methods more detailed and with higher confidence than earlier. As a part of the COST activity, a benchmark dataset was built whose characteristics approach well the characteristics of real networks of observed time series. This dataset offers much better opportunity than ever before to test the wide variety of homogenisation methods, and analyse the real effects of selected theoretical recommendations.

Empirical results show that real observed time series usually include several inhomogeneities of different sizes. Small inhomogeneities often have similar statistical characteristics than natural changes caused by climatic variability, thus the pure application of the classic theory that change-points of observed time series can be found and corrected one-by-one is impossible. However, after homogenisation the linear trends, seasonal changes and long-term fluctuations of time series are usually much closer to the reality than in raw time series. Some problems around detecting multiple structures of inhomogeneities, as well as that of time series comparisons within homogenisation procedures are discussed briefly in the study.

1 Introduction

To obtain a precise and reliable picture about the climatic variability of the period with instrumental observation methods, it is necessary to eliminate the influence of technical changes (hereafter: inhomogeneity, IH) in the observation systems. Therefore, together with the collection and archiving the observational data, a special branch of quality control developed for managing this kind of problem, i.e. the so-called time series homogenisation. During this development, a large number of statistical methods were introduced. Recently, enhanced efforts have been devoted to compare and evaluate the efficiency of different methods and this is not an easy task, because in real observed datasets the true statistical properties of IHs are never known exactly. Among other efforts the COST HOME has brought dynamism to these examinations. The present epoch of research on homogenisation methods can be characterized with the following new lines:

(i) Homogenisation methods are tested in simulated databases whose properties approach well the real properties of networks of observed climatic time series; (ii) The performance of homogenisation methods is evaluated by calculating RMSE between corrected time series and the corresponding homogeneous time series, as well as calculating the mean bias of linear trends between corrected time series and perfect time series.

In this study some theoretical problems related to the application of homogenisation methods are briefly described, and an example is shown for demonstrating the superior performance of the detection methods whose algorithm includes a direct identification of multiple structures of change-points.

2 Methods and definitions

Two, frequently appearing forms of IHs are defined here. Note that other forms may also occur, but they are not discussed in this study.

- *Change-point*: A sudden shift in the mean of the observational values. It is the most frequent form of IH, since most technical changes happen abruptly.

– *Platform-like inhomogeneity* [Pfm]: Pair of change-points of the same size, but of the opposite direction.

Concepts related to efficiency-evaluation:

– *Correct detection*: When an IH is detected in year j, and an IH really exists in section $[j-2, j+2]$.

– *False detection*: In the detection result an IH is included in year j, and no IH exists in section $[j-2, j+2]$. Note: If two IHs are detected around a really existing IH (e.g. an IH really exists in year j, but the detection results indicate two IHs, one in $j-2$ and another one in $j+2$) one of them is sorted into the correct detections, but the other one into the false detections.

The total number of correct detections, that of false detections and that of true IHs are denoted by S_R, S_F and S, respectively.

– *Power of detection* (Pw):

$$Pw = \frac{S_R}{S} \qquad (1)$$

In Eq. (1), S stands for the total number of true IHs.

– *Detection skill* (D):

$$D = \frac{S_R - S_F}{S} \qquad (2)$$

– *Efficiency of RMSE-reduction* (E): In this study the efficiency is characterised by the improvement of root mean squared error (RMSE) due to homogenisation.

$$E = \frac{RMSE_{raw} - RMSE_{homogenised}}{RMSE_{raw}} \qquad (3)$$

3 Purpose of homogenisation

Until the recent years it was common that the efficiency of homogenisation methods was evaluated by some set of simulated time series including a white noise process plus one or a few change-points whose magnitudes are significantly higher than the standard deviation of noise (s_d). In traditional evaluations of usability, most often the detection skill was calculated only. Although this kind of examinations provide valuable information about the properties of homogenisation methods, the results do not give direct information about the effectiveness of the methods, for three reasons: (i) Real properties of observed climatic time series are very different from this simple model, (ii) In the calculation of D the use of some arbitrary parameters is unavoidable, (iii) Detection skill, hit rate, false alarm rate, etc. do not provide direct information about the success in improving the reliability of trends and long-term fluctuations in time series. These ideas had already been taken into account (though in a relatively initial

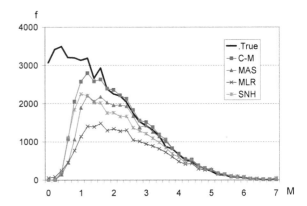

Figure 1. Magnitude-distribution of real and detected IHs (change-points) when the mean frequency of change-points is 5 per 100 yr, and shift-magnitudes have normal distribution with 0 peak and 3.5 times larger standard deviation than s_d. On the abscissa M means magnitude proportioned to s_d, while frequencies (f) are shown with an arbitrary unit. Homogenisation methods: C-M – PRODIGE, MAS – Multiple Analysis of Series for Homogenization, MLR – Multiple Linear Regression, SNH – Standard Normal Homogeneity Test.

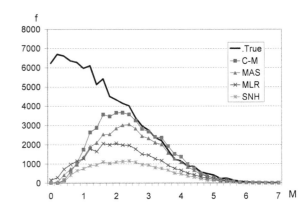

Figure 2. The same as Fig. 1, but 5 Pfm (i.e. 10 change-points of platform-like IHs) per 100 yr are included in time series instead of persistent shifts. Length of platforms has equal distribution between 1 month and 10 yr.

level) in some studies before the COST HOME. Menne and Williams (2005, 2009) analysed the efficiency of homogenisation methods applying test dataset in which the sizes of IHs have normal distribution with zero peak. Domonkos (2008, 2011) built a test dataset whose statistical characteristics are very close to an observed temperature dataset and used various measures for assessing efficiency.

After examining the homogeneity of radiosonde time series, Sherwood (2007) stated: "...detection of change-points is neither realistic nor necessary, ...success should be measured instead by the integrity of climate signals." This thesis is valid also for time series of surface observations. The statistical properties of true and detected IHs often differ markedly. In Figs. 1–2 some experimental results are shown that obtained by examining sets of 10 000 artificial time series including 5 change-points (Fig. 1), or 5 Pfms (Fig. 2). When the sizes of IHs are relatively large, the amount of detected IHs approaches well their true frequency, but for small IHs detection is often impossible by any of the known homogenisation methods. The ratio of undetected IHs is considerably higher when Pfms occur in time series.

Thinking over Sherwood's thesis it can be stated that time series homogenisation is the utilisation of the spatially redundant information for the improvement of reliability of time-variability in data. (Note that instead of or beside the spatial redundancy, other pieces of information or assumptions can also be used for homogenisation, but in climatological studies it is not typical and usually not recommendable.) Characteristics of effective homogenisation methods are: (i) maximal exploitation of spatial information, (ii) high skill in finding timings of IHs, (iii) ability to treat common effects of multiple IHs, (iv) application of an appropriate correction method.

In testing efficiency the most reliable results can be obtained when statistical properties of artificial datasets are close to those of observed datasets. For obtaining test datasets with realistic properties, characteristics of detected IHs should be compared between observed data and artificial data (Domonkos, 2008, 2011). The results of this kind of tests indicate that in observed climatic time series small shifts and Pfms are frequent, thus their direct and indirect effects must be taken into account in the evaluation of homogenisation methods.

4 Common effects of multiple inhomogeneities

In most homogenisation methods used in climatic studies (Standard Normal Homogeneity Test (SNHT), Alexandersson, 1986; Multiple Linear Regression (MLR), Vincent, 1998; Penalised Maximal t-test (PMT), Wang et al., 2007, etc.) a step-by-step procedure is applied, in which methods detect only one IH in a particular step, thereafter time series are cut into two parts at the timing of the detected change-point. This cutting algorithm can be transformed to semi-hierarchic algorithm with supplying the procedure with some other steps (Lanzante, 1996; Moberg and Alexandersson, 1997, etc.), but experimental results indicate that semi-hierarchic algorithms do not provide substantial improvement relative to the cutting algorithm (Domonkos, 2011). Some methods detect multiple structures of IHs in a direct way (Multiple Analysis of Series for Homogenisation, Szentimrey, 1999; PRODIGE, Caussinus and Mestre, 2004, these are referred as "direct methods" hereafter). When the number of IHs is low, cutting algorithm and semi-hierarchic algorithm may function efficiently (Menne and Williams, 2005), but for complex structures of IHs only the direct methods are powerful (Domonkos, 2011).

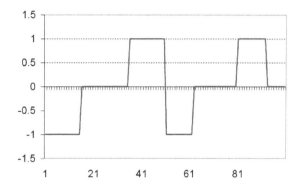

Figure 3. An example for complex structures of IHs. The horizontal axis is for the serial numbers of years, while in the vertical axis cumulative effects of IHs are shown. Noise is excluded for the sake of demonstrativeness.

To demonstrate the limitations of hierarchic methods, a theoretical example is presented here, i.e. a particular structure of IHs without noise (Fig. 3). In this example the largest shift is just in the middle of the time series and the means for the two halves of the time series are the same. For this kind of IH-structure, hierarchic algorithms are often incapable of detecting the largest IH in the first step, particularly when significance-examinations are based on the step-by-step comparison of some statistical characteristics between two parts of the series (SNHT, PMT, extremes of accumulated anomalies, as well as non-parametric methods, as for instance the Wilcoxon Rank Sum Test (WRS), Wilcoxon, 1945). However, a failure in finding the largest IH in the first step might affect the final results of the homogenisation procedure in hierarchic methods, since (i) the other (smaller) IHs can be detected with relatively low certainty because of their small size, (ii) in hierarchic algorithms a possible error in the first step introduces bias for the initial condition of later steps.

The performance of six detection methods (Easterling and Peterson method (E-P, Easterling and Peterson, 1995), MASH, MLR, PRODIGE, SNHT, WRS) is analysed by supplying the IH-structure of Fig. 3 with standard white noise in 10 000 simulation experiments. Detection powers for the different methods in function of the IH-size relative to s_d are calculated for the largest IH in the middle of the time series. The results are shown in Fig. 4, and it can be seen that the direct methods are really more effective than the hierarchic methods in identifying the largest IH of the time series. On the other hand, E-P has an even better Pw, than the direct methods. Note that E-P does not belong to the direct methods, nor to the hierarchic methods. In Fig. 5 the detection skills are shown for the same methods and IH-structures, as they included in Fig. 4. In the calculation of D all the IHs of the time series were taken into account. The results show that for small IHs the E-P still has the best performance, but when the IH sizes are larger than the background noise, the

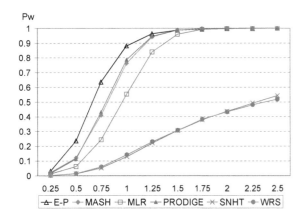

Figure 4. Power of detection for the largest IH of Fig. 3. On the abscissa the unit is s_d.

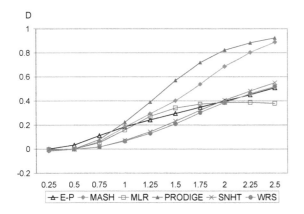

Figure 5. Detection skill for all IHs of Fig. 3. On the abscissa the unit is s_d.

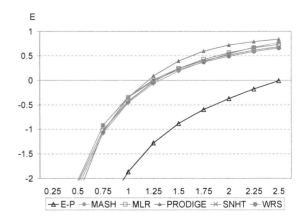

Figure 6. Efficiency in reducing RMSE when the IH-structure is the same as in Fig. 3. On the abscissa the unit is s_d.

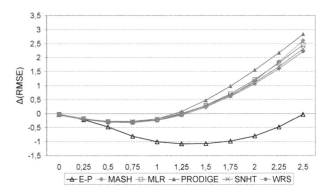

Figure 7. The same as Fig. 6, but the improvement of RMSE-error is presented on an absolute scale. $\Delta(RMSE) = RMSE_{raw} - RMSE_{homogenised}$.

PRODIGE and MASH perform best. It is interesting to see that the MLR which has relatively good skill in detecting the largest IH (because the significance test of MLR is based on the autocorrelation of the tested series and not on the comparison of characteristics between two parts of the series), loses this advantage relative to SNHT and WRS when all the detection results are evaluated together. Figure 6 shows the efficiencies in the improvement of RMSE. These results have two striking features, i.e. (a) in case of small-size IHs all the efficiencies are markedly negative, (b) one of the six methods examined, namely the E-P, always performs with negative efficiency when the IH-sizes are lower than 2.5. The latter is the consequence of the fact that in E-P the time-coherence between the pieces of the detection results is less organised than in an hierarchic method (and even less organised than in direct methods). The results of E-P provide clear evidence that the examination of detection skill (detection power, false alarm rate, etc.) is not satisfactory in itself to evaluate the efficiency of homogenisation methods. On the other hand, the negative efficiencies for low-size IHs must not discourage users from applying homogenisation methods: The RMSE error in raw time series of 0.5 characteristic IH-size is only

4 % of the RMSE in raw time series with IHs of size 2.5. Thus, in absolute scale (Fig. 7), time series with low-size IHs (or without IH) might suffer a little corruption, while the quality of time series with large IHs can be improved markedly by the application of homogenisation methods. In the present experiment, PRODIGE performs best for IH sizes of larger than 1.25, while there is very little difference among the efficiencies of MASH, MLR, SNHT and WRS. Other experiments (Domonkos, 2011) confirm that (a) PRODIGE performs best when large- or medium-size IHs occur in time series, (b) PRODIGE and MASH have markedly better detection skills than other methods when complicated IH-structures occur in time series, (c) PRODIGE usually has superior performance in estimating linear trends with homogenised time series, (d) methods that do not consider the connections between the pieces of the detection results for individual IHs, either by direct algorithms or hierarchic algorithms, cannot be recommended for homogenising climatic time series, because the resultant time series often contain large errors, even if the detection skills of the methods are good.

5 Spatial comparison of time series

Since rapid temporal changes in climatic time series might occur also by true climatic variability, homogenisation methods are usually applied to differences of the raw time series. By generating these series the impact of climatic variability is reduced, because it is common for a given climatic region. However, a general problem of the spatial comparisons is that IH-detection results might be affected by the IHs in the series with which the candidate series are compared. Ideally, homogeneous reference series should be found for each candidate series. As this expectation is unrealistic, often a series of pair-wise comparisons is recommended instead of using fixed reference series (e.g. Caussinus and Mestre, 2004). In pair-wise comparisons, change-points existing in any time series of the network are searched and treated individually. However, pair-wise comparison methods also have drawbacks: (i) They use restricted number of series from the neighbourhood (i.e. in one particular comparison only one time series), thus noise and undetected IHs might cause relatively large errors; (ii) Skilled algorithms of multiple pair-wise comparisons can be too complicated for applying them in automatic procedures.

In recent efficiency examinations using the benchmark dataset of COST HOME, it turned out that a traditional creation of reference series may provide competitive efficiency. The method ACMANT (Domonkos et al., 2011) which is a modified and automated version of PRODIGE, performed with efficiency similar to the best of other methods (PRODIGE and MASH, Venema et al., 2010). The role of individual IHs in composites of reference series declines with the increase of the number of the composites. This fact is exploited in ACMANT in the way that the reference series is the weighted average of surrounding time series, as it was recommended by Peterson and Easterling (1994). In the present version of ACMANT the number of reference-composites is unlimited, and the minimum threshold of acceptable spatial correlation is 0.4. Notwithstanding, the author thinks that the optimal way of spatial comparison needs much further examination, because it is hard to find the optimal combination of the following two competitive aspects: On the one hand, impacts of climatic differences (which is indicated by relatively low spatial correlation) should be excluded by using a limited number of composites in building reference series, but, on the other hand, impacts of undetected IHs and noise should be reduced by including as many composites as possible, since with the use of a larger number of composites the effects of individual errors in composites decrease. Naturally, if effective and user-friendly versions of pair-wise comparisons are available, their use will also be recommendable.

6 Discussion and conclusions

The developers and users of homogenisation methods have to bear in mind that the eventual purpose of homogenisation is not to find change-points, but to obtain an improvement in the quality of the observational datasets that gives the opportunity to achieve more precise and more reliable results in climate change and climate variability analyses. Some old rules and recommendations should be re-evaluated. For instance, the performance of homogenisation methods depends on the connections between the pieces of the detection result, thus individual subjective decisions for selected change-points (e.g. using metadata information) may introduce undesired uncertainty to the overall efficiency of the procedure. Further examinations are needed also to find the optimal way of spatial comparison. As examples show, pair-wise comparison technique, but also the classic way of building reference series may both work with high efficiency within some homogenisation procedures.

The selection of the best homogenisation methods has to be based on efficiency tests executed on artificial databases of climatic time series with realistic statistical properties. We should go further on the way that is marked by the COST HOME activity.

Edited by: M. Brunet-India
Reviewed by: B. Trewin and two other anonymous referees

References

Alexandersson, H.: A homogeneity test applied to precipitation data, J. Climatol., 6, 661–675, 1986.

Caussinus, H. and Mestre, O.: Detection and correction of artificial shifts in climate series, J. Roy. Stat. Soc. Series, C53, 405–425, 2004.

Domonkos, P.: Testing of homogenisation methods: purposes, tools and problems of implementation, Proceedings of the 5th Seminar and Quality Control in Climatological Databases, edited by: Lakatos, M., Szentimrey, T., Bihari, Z., and Szalai, S., WCDMP-No. 71, WMO/TD-NO. 1493, 126–145, 2008.

Domonkos, P.: Efficiency evaluation for detecting inhomogeneities by objective homogenisation methods, Theor. Appl. Climatol., doi:10.1007/s00704-011-0399-7, 2011.

Domonkos, P., Poza, R., and Efthymiadis, D.: Newest developments of ACMANT, Adv. Sci. Res., 6, 7–11, doi:10.5194/asr-6-7-2011, 2011.

Easterling, D. R. and Peterson, T. C.: A new method for detecting undocumented discontinuities in climatological time series, Int. J. Climatol., 15, 369–377, 1995.

Lanzante, J. R.: Resistant, robust and non-parametric techniques for the analysis of climate data: theory and examples, including applications to historical radiosonde station data, Int. J. Climatol., 16, 1197–1226, 1996.

Menne, M. J. and Williams Jr., C. N.: Detection of undocumented changepoints using multiple test statistics and composite reference series, J. Climate, 18, 4271–4286, doi:10.1175/JCLI3524.1, 2005.

Menne, M. J. and Williams Jr., C. N.: Homogenization of temperature series via pairwise comparisons, J. Climate, 22, 1700–1717, doi:10.1175/2008JCLI2263.1, 2009.

Moberg, A. and Alexandersson, H.: Homogenization of Swedish temperature data. Part II: Homogenized gridded air temperature compared with a subset of global gridded air temperature since 1861, Int. J. Climatol., 17, 35–54, 1997.

Peterson, T. C. and Easterling, D. R.: Creation of homogeneous composite climatological reference series, Int. J. Climatol., 14, 671–679, 1994.

Sherwood, S. C.: Simultaneous detection of climate change and observing biases in a network with incomplete sampling, J. Climate, 20, 4047–4062, doi:10.1175/JCLI4215.1, 2007.

Szentimrey, T.: Multiple Analysis of Series for Homogenization (MASH), WCDMP-41, WMO-TD. 932, Geneva, 27–46, 1999.

Venema, V., Mestre, O., and the COST HOME Team: Benchmark database, EGU General Assembly, Vienna, Austria, 3–7 May 2010, EGU CL4.6-13357, 2010.

Vincent, L. A.: A technique for the identification of inhomogeneities in Canadian temperature series, J. Climate, 11, 1094–1104, 1998.

Wang, X. L., Wen, Q. H., and Wu, Y.: Penalized maximal t test for detecting undocumented mean change in climate data series, J. Appl. Meteor. Climatol., 46, 916–931, doi:10.1175/JAM2504.1, 2007.

Wilcoxon, F.: Individual comparisons by ranking methods, Biometrics Bull., 1, 80–83, 1945.

9

1961–1990 monthly high-resolution solar radiation climatologies for Italy

J. Spinoni[1,*], **M. Brunetti**[2], **M. Maugeri**[1,2], **and C. Simolo**[2]

[1]Department of Physics, University of Milan, Milan, Italy
[2]Institute of Atmospheric Sciences and Climate, National Research Council (ISAC-CNR), Bologna, Italy
[*]now at: JRC-IES, Ispra, Italy

Correspondence to: J. Spinoni (jonathan.spinoni@jrc.ec.europa.eu)

Abstract. We present a methodology for estimating solar radiation climatologies from a sparse network of global radiation and/or sunshine duration records: it allows to obtain high-resolution grids of monthly normal values for global radiation (and for the direct and diffuse components), atmospheric turbidity, and surface absorbed radiation. We discuss the application of the methodology to a preliminary version of an Italian global radiation and sunshine duration data set, which completion is still in progress and present the resulting 1961–1990 monthly radiation climatologies.

1   Introduction

High-resolution datasets of monthly climatological normals (i.e. high-resolution climatologies) have proved to be increasingly important in the recent past, and they are likely to become even more important in the future. They are used in a variety of models and decision support tools in a wide spectrum of fields such as, just to cite a few, energy, agriculture, engineering, hydrology, ecology and natural resource conservation (Daly et al., 2002; Daly, 2006).

One of the most important variables for a lot of possible applications is solar radiation. Even though some examples of solar radiation climatologies are already available for Italy (see e.g. Lavagnini and Jibril, 1991; Petrarca et al., 2000; Suri and Hofierka, 2004), they suffer the lack of a dense network of long records of observational data.

In this context we set up a research program with the aim of (i) setting up an extensive data base of Italian global radiation and sunshine duration records and (ii) developing a methodology for estimating high resolution solar radiation climatologies from these records. Sunshine duration records have the great advantage, with respect to global radiation records, of a much larger data availability, especially when long-term records are considered.

The methodology for estimating solar radiation climatologies from global radiation and/or sunshine duration records has been developed within a Ph.D. thesis recently concluded at Milan University (Spinoni, 2010). It consists in (i) con-

verting sunshine duration data into global radiation (if global radiation data are not available for the site), (ii) decomposing global radiation into a direct and a diffuse component, (iii) gridding direct and diffuse components of global radiation, (iv) evaluating atmospheric turbidity over the same grid by means of the direct component of global radiation, (v) calculating direct, diffuse and reflected components of global radiation for any cell of the used grid, taking into account its slope and aspect and considering shading, (vi) calculating the corresponding absorbed radiation by means of land-use-based albedo estimations. The first application of the methodology consisted in the estimation of global radiation climatologies (Spinoni, 2010) that have been used as proxies to support the construction of 1961–1990 temperature climatologies for Italy. For this application we were mainly interested to the result produced at the last point. Other applications, however, might require the results produced at the other points or modified versions of them (e.g. for solar energy production it might be interesting to produce the result of point v, substituting the slope and the aspect of the grid-cell, with values corresponding to an hypothetical panel in a solar plant).

The paper aims at presenting this methodology and showing a preliminary version of radiation climatologies obtained applying it to the Italian global radiation and sunshine duration records that are already available in digital form. The climatologies are represented on the USGS GTOPO30 Digital Elevation Model grid (USGS, 1996), i.e. with a

30 arc-second resolution, corresponding to about 700 m in the W-E direction and about 900 m in the S-N direction. This DEM has also been used to estimate the slope and the aspect of the surface and the rate of shading due to the surrounding areas.

2 Sunshine and solar radiation data

The activities aiming at setting up an extensive data base of Italian sunshine duration records are in progress. They will include the digitisation of a great amount of data that are available only on paper forms, allowing to greatly extend the records that are available at present time. The result will not only be used for describing the spatial distribution of solar radiation normal values, but will also allow the temporal trends over different climatic regions of Italy to be studied.

The new data set is however not available yet. So, we started to develop our methodology with a more limited data availability. In particular we considered monthly records which are already available in digital form from Italian Air Force, ENAV (www.enav.it), CRA-CMA (http://www.cra-cma.it), some regional Environmental Agencies and the European Solar Radiation Atlas (Kasten et al., 1984). The standard method for sunshine duration observation was the Campbell-Stokes instrument; we have however not yet full information on the measures as the data and metadata collection is still in progress.

A significant part of the data used in this paper is available from the web (see http://www.scia.sinanet.apat.it/ and http://www.cra-cma.it/). The main deficit of these data is that many records cover rather short time periods and only a minor fraction of them has no missing data in the 1961–1990 period. So the monthly station climatic normals which can be obtained from these records are not completely representative of this period. This problem may introduce a significant bias as it is well demonstrated that sunshine duration and solar radiation are not constant through decades: records show a "global early brightening" period approximately between 1940 and 1950, a "global dimming" period between 1950 and 1980 and a "global late brightening" period after 1980 (Wild et al., 2005; Ohmura, 2006; Wild, 2009). Furthermore, solar radiation and sunshine duration are influenced by great volcanic explosions such as, e.g. Eyjafjallajökull (Iceland, 2010). Due to these and other facts, which cause variability over a wide range of time scales (see e.g. Brunetti et al., 2009), sunshine duration data that are not related to the reference period should be handled with care. A complete solution to this problem will be possible only when the full data set will become available, as the conversion of all station normal values to 1961–1990 normals requires the knowledge of the spatial distribution of the time evolution of sunshine duration over Italy. At present time we only fixed a minimum length of the records (10 yr) and performed some preliminary comparisons among neighbouring stations in order to

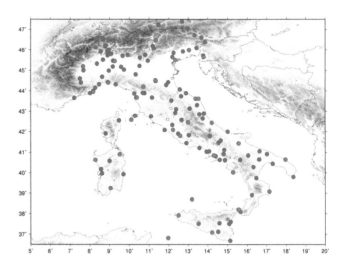

Figure 1. Stations with sunshine duration records. Red dots: also global radiation available.

exclude from the analyses the records which seem to exhibit the largest bias due to missing data in the 1961–1990 period.

Besides the records from the listed sources, we considered also monthly 1961–1990 normals from a monographic book (Cat Berro et al., 2005) and from the web for a few more Italian stations and for about 30 stations of the surrounding countries (France, Switzerland, Austria and Slovenia).

The final data set used in this paper consists of 158 stations (see Fig. 1). For all stations, being they declared at WMO standard, we assume they are far away from surrounding shading obstacles. For 31 of the stations, besides the sunshine duration records, we also use global radiation 1961–1990 monthly normals. These stations are from the Italian Air Force network (see www.meteoam.it).

3 Estimating flat surface solar radiation grids from station data

3.1 Converting sunshine duration to global radiation

The first step of our procedure consists in estimating the monthly clearness index (K_t) values for all the stations of our data set that do not have global radiation data. The clearness index is the ratio between the global radiation received by a surface (H_T) and the exo-atmospheric radiation received by the same surface (H_0) (Gueymard, 2001). K_t includes cloudiness and turbidity. Following the Angstrom-Black's equation (Black et al., 1954), it can be estimated from relative sunshine duration (i.e. from the ratio between the number of sun hours measured by a sunshine recorders (S) and the solar day length from sunrise to sunset (S_0)):

$$K_t = \frac{H_T}{H_0} = a + b\frac{S}{S_0} \qquad (1)$$

In Eq. (1) we consider sunshine duration as representative of the 15th day of each month and we use corresponding values

Table 1. Monthly a, b coefficients in Eq. (1) and corresponding standard errors based on the 31 Italian Air Force station with both global radiation and sunshine duration data.

Month	$a \pm \sigma_a$	$b \pm \sigma_b$
January	0.16 ± 0.02	0.69 ± 0.06
February	0.17 ± 0.03	0.71 ± 0.08
March	0.21 ± 0.03	0.65 ± 0.06
April	0.18 ± 0.03	0.72 ± 0.06
May	0.18 ± 0.03	0.69 ± 0.05
June	0.18 ± 0.03	0.68 ± 0.05
July	0.21 ± 0.05	0.61 ± 0.05
August	0.22 ± 0.03	0.58 ± 0.05
September	0.27 ± 0.03	0.52 ± 0.05
October	0.22 ± 0.02	0.60 ± 0.04
November	0.18 ± 0.02	0.65 ± 0.04
December	0.17 ± 0.02	0.67 ± 0.06

for S_0 and H_0: they only depend on astronomical and geographical factors and they can be calculated according to standard procedures (see Spinoni, 2010 for details).

Actually, before using Eq. (1) for estimating the monthly clearness indexes of all our 158 sites, we use the sites with global radiation data to estimate the coefficients a and b. The results we obtain with present time data availability (31 stations) are shown in Table 1, together with their standard errors. Beside these coefficients, 8 other alternative sets of coefficients found in the literature (Rietveld, 1978; Landsberg, 1981; Andretta et al., 1982; Iqbal, 1983; Newland, 1989; Gopinathan and Soler, 1995; Akinoglu and Ecevit, 1990; Coppolino, 1994) were tested. The errors turned however out to be larger than the ones obtained with the coefficients tuned on Italian data and so we prefer using these coefficients. The Mean Absolute Error (MAE) over all months and stations gives a synthetic information on the ability of Eq. (1) to get the observed clearness index: it results 0.021 with our coefficients whereas it ranges between 0.025 and 0.108 with the 8 other alternative sets of coefficients. When the full dataset will be available, the estimation of the clearness index from sunshine duration data has to be studied more in detail by investigating the use of local coefficients (see e.g. Scharmer et al., 2000) and by trying to consider other variables beside sunshine duration.

Once the monthly clearness index values are available, the corresponding global radiation normals can simply be calculated as:

$$H_T = K_t H_0 \tag{2}$$

3.2 Decomposition models: from global radiation to direct and diffuse radiation

After global radiation normals are available for all stations, we estimate the direct and diffuse components of solar radia-

tion by means of the so called decomposition models (we assume that our sunshine duration data, being measured under WMO standard conditions, are not influenced by reflected radiation). In particular, we use a decomposition model based on Eq. (3) (Iqbal, 1983).

$$K_{dif} = \frac{H_{dif}}{H_T}, \quad K_{dir} = \frac{H_{dir}}{H_T}, \quad K_{dif} + K_{dir} = 1 \tag{3}$$

K_{dif} is the diffuse radiation fraction of the global radiation received by a surface, K_{dir} is the corresponding direct radiation fraction, H_{dif} and H_{dir} are the diffuse and direct components of global radiation.

Decomposition models should be based on local data. However, when local data are not available, models which are reasonably valid worldwide can be used, e.g. the third order polynomial model used in the European solar Radiation Atlas (Erbs et al., 1982; Scharmer et al., 2000) or the models proposed by Page (1964); Iqbal (1983); Reindl et al. (1990) and Gopinathan and Soler (1995). In our methodology we use (Spinoni, 2010) the following relation (Gopinathan and Soler, 1995):

$$K_{dif} = 0.878 - 0.3328\, K_t - 0.53\frac{S}{S_0} \tag{4}$$

Therefore, in the second step of our procedure, we calculate, by means of Eqs. (3)–(4), the monthly diffuse and direct fractions for all our stations.

3.3 Gridding of direct, diffuse, and global radiation for flat surfaces

The third step of our procedure consists in using, for each month, direct and diffuse components of station global radiation to construct high-resolution grids covering all the Italian territory. This gridding procedure is performed, on each node of the USGS GTOPO30 DEM, by means of an Inverse Distance Gaussian Weighting (IDGW) spatialisation model, using Gaussian radial weights (w_i^{rad}) for the contribution of each station:

$$w_i^{rad}(x,y) = \exp\left(-\frac{d_i^2(x,y)}{c_d}\right) \tag{5}$$

where $d_i(x,y)$ is the distance between the i-th station and the considered grid-cell and c_d is a coefficient regulating the decrease of the weighting factor with distance: it is chosen in order to have weight equal to 0.5 at distance $\bar{d}$:

$$c_d = -\frac{\bar{d}^2}{\ln(0.5)} \tag{6}$$

On the basis of present time data availability, we choose $\bar{d} = 50\,\text{km}$.

When the full dataset will be available, we will also use more complex spatialisation techniques, trying to take into account the effect of geographic variables.

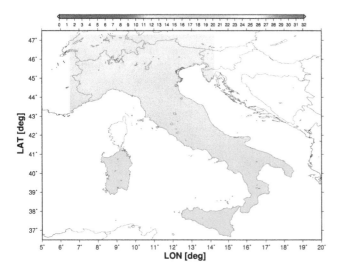

Figure 2. 1961–1990 yearly average global radiation MJ/day $\times$ m^2 on flat surfaces.

The gridding procedure allows obtaining monthly high-resolution fields of direct, diffuse and global radiation that are representative of flat and non-shaded surfaces. Figure 2 shows, as an example, the yearly average global radiation that we obtain for such surfaces with present time data availability.

4 Evaluation of the turbidity of the atmosphere

The fourth step of our procedure consists in evaluating the spatial distribution of atmospheric turbidity. This evaluation is based on the following relation (Iqbal, 1983):

$$H_{\text{dir}} = E_0 \, I_0 \left(\int_{\text{sunrise}}^{\text{sunset}} dh \, \cos(\theta_{\text{inc}}) \exp[T_{\text{F}} \, m_{\text{A}} \, \delta_{\text{R}}(m_{\text{A}})] \right) \quad (7)$$

where H_{dir} is the direct component of global radiation calculated for the 15 day of each month as described in the previous sections, E_0 is the eccentricity factor (i.e. the correction due to the elliptical orbit of the Earth), I_0 is the solar constant, θ_{inc} is the solar angle of incidence and the exponential part explains the attenuation due to the atmosphere: T_{F} is the turbidity factor, m_{A} is the optical air mass, δ_{R} is the Rayleigh's depth of the atmosphere. T_{F} represents the turbidity of the vertical column of the atmosphere over the grid cell: clouds, water vapor, pollution, fog, ozone, and many other factors are included in T_{F}.

For each point and each month we search for the T_{F} best matching the H_{dir} in Eq. (7). We consider for the integration 5-min time intervals (dh) and calculate the time dependent variables (θ_{inc}, m_{A} and δ_{R}) over each interval. In the calculation of m_{A} and δ_{R} we also consider elevation and take into account the refraction of the atmosphere (see Kasten and Young, 1989; Rigollier et al., 2000). Details on the calculations can be found in Spinoni (2010). This step of the proce-

dure allows obtaining monthly atmospheric turbidity 1961–1990 normals over the same grid used to spatialise H_{dir}.

5 Solar radiation model for inclined surfaces

Once we know the turbidity of the atmosphere, we can calculate the solar radiation received by inclined surfaces. This calculation requires the knowledge of the slope and the aspect of each grid-cell, as well as the evaluation of the shading due to the surrounding grid-cells: this information is obtained by means of the GTOPO30 DEM.

Direct radiation for inclined surfaces ($H_{\text{dir}}^{\text{incl}}$) is calculated with a slightly modified version of Eq. (7), i.e. introducing a binary factor J that represents shading: it is obtained by exploring the grid-cells surrounding each node of the GTOPO30 DEM and checking, with a 5-min time resolution, if the path from the node to the sun does or does not intercept the DEM surface. If the grid cell is shadowed in the 5-min interval that we use in the integration, J is set to 0, otherwise it is set to 1. In this case θ_{inc} is naturally calculated taking into account the slope and the aspect of the surface.

$$H_{\text{dir}}^{\text{incl}} = E_0 \, I_0 \left(\int_{\text{sunrise}}^{\text{sunset}} dh \, J \, \cos(\theta_{\text{inc}}) \exp[T_{\text{F}} \, m_{\text{A}} \, \delta_{\text{R}}(m_{\text{A}})] \right) \quad (8)$$

Actually, in spite of the analogies of Eqs. (7) and (8), they are used in a completely different way: in fact in Eq. (7), we know H_{dir} and use it to get T_{F}; on the contrary, in Eq. (8) we know T_{F} and use it to get $H_{\text{dir}}^{\text{incl}}$.

Diffuse radiation for inclined surfaces ($H_{\text{dif}}^{\text{incl}}$) is calculated considering diffuse radiation as isotropic (as it is usual in solar radiation models made for climate-related purposes). In order to obtain the grids, we just multiply the diffuse radiation received by a flat surface by the sky view factor (V_{F}), i.e. the visible fraction of the sky from the grid-cell. In our procedure this factor is assumed to be dependent only on the slope (s) of the grid-cell itself. More details on this assumption can be found in Chung and Yun (2004).

$$H_{\text{dif}}^{\text{incl}} = H_{\text{dif}} \, V_{\text{F}}, \quad V_{\text{F}} = \frac{1 + \cos(s)}{2} \quad (9)$$

Reflected radiation for inclined surfaces ($H_{\text{ref}}^{\text{incl}}$) is calculated as:

$$H_{\text{ref}}^{\text{incl}} = (H_{\text{dir}}^{\text{incl}} + H_{\text{dif}}^{\text{incl}}) \, O_{\text{SF}} \, \alpha, \quad O_{\text{SF}} = 1 - V_{\text{F}} \quad (10)$$

where O_{SF} is the obstructed sky factor (see Chung and Yun, 2004), i.e. the obstructed portion of the sky and α is the ground albedo. In our procedure we assume for the albedo the value which we attribute to the grid-cell itself, even though the reflection is due to the surrounding cells. This approach is justified as the very limited contribution of reflected radiation that we have with a DEM resolution of 30 arc-seconds, does not justify the much greater complexity which would be necessary in order to take into account the slope, aspect and albedo of the surrounding grid-cells.

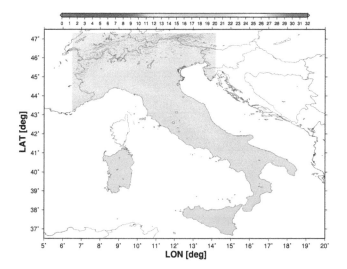

Figure 3. 1961–1990 yearly average global radiation MJ/day$\times$m^2.

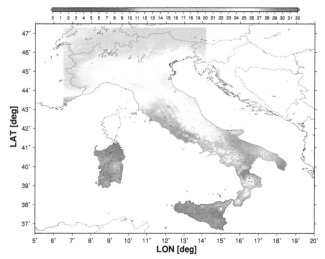

Figure 5. As in Fig. 3, but for July.

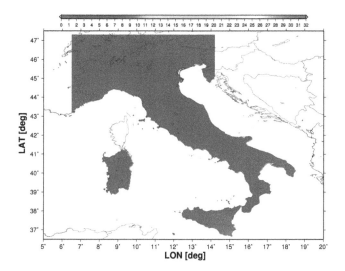

Figure 4. As in Fig. 3, but for January.

The final results are 1961–1990 monthly climatologies for global and absorbed radiation. Some examples are shown in Figs. 3–5.

6 Validation

We performed a validation of the preliminary climatologies that we present in this paper. This validation is based on a subset of 28 of the Air Force stations with 1961–1990 global radiation normals: they were selected focusing only on stations located in non inclined grid-cells. In other terms we require, not only that a station is located on flat surface, but also that the grid-cell in which it is located is completely flat.

The agreement between the climatologies and the station data was evaluated by means of the mean absolute error (MAE) and the relative mean absolute error (MAER). Due to the fact that we used the same data set to evaluate the clearness index, we performed a leave-one-out validation, removing from the input data the station that is, in turn, evaluated. The results are shown in Table 2: MAERs are smaller in late winter and spring than in autumn and early winter, but no systematic over or underestimation was found. The average MAER is under the threshold of 5 %, but such a validation is based on a very small data set and has to be considered as preliminary.

7 Conclusions and area for further work

We described a methodology for the construction of solar radiation and atmospheric turbidity normal value grids. It requires solar radiation or sunshine duration data. All the points of the methodology described in the paper have been encoded in an unique program which allows the user to handle the different steps of the calculations.

Albedo is estimated by means of the GLC2000 land cover grid provided by Joint Research Center (see the website: http://bioval.jrc.ec.europa.eu/products/glc2000/glc2000.php) and on the basis of literature albedo-cloud cover relations (see e.g. Hummel and Reck, 1979; Henderson-Sellers and Wilson, 1983; Wilson and Henderson-Sellers, 1985). Albedo is not corrected for fresh snow cover in winter, because snow climatologies for Italy are not available.

Once direct, diffuse and reflected radiation are available, we simply calculate global radiation summing them. All the procedure is naturally performed for all the grid-points of Italy. The absorbed radiation is then obtained simply by considering the albedo factor.

$$H_{glob}^{incl} = (H_{dir}^{incl} + H_{dif}^{incl} + H_{ref}^{incl}), \quad H_{abs}^{incl} = H_{glob}^{incl}(1-\alpha) \qquad (11)$$

Table 2. Estimated global radiation MAE (MJ/day × m^2) and MAER (%) evaluated comparing the modeled climatologies with the station 1961–1990 normals.

Month	MAE (MJ/day × m^2)	MAER (%)
January	0.27	4.4
February	0.31	3.5
March	0.45	3.4
April	0.86	4.9
May	0.78	3.6
June	1.05	4.4
July	1.16	4.8
August	0.97	4.6
September	0.81	5.0
October	0.55	5.1
November	0.36	5.3
December	0.33	6.3
Year	0.66	4.6

Several points require future improvements. The first point on which we are already working consists in a significant enlargement of the data set: it will allow both to enlarge the number of records and to extend the time coverage of each record; moreover it will allow to make available, beside long-term sunshine records, also shorter radiation records that will be used to better study the relation between global radiation and sunshine duration. Another important point that will be considered in the future consists in the use of a DEM with higher resolution: this will improve the evaluation of the influence of the grid-cells surrounding each grid-cell: they regulate shading and influence also direct radiation (by means of the sky view factor) and reflected radiation.

Acknowledgements. We sincerely thank all the data providers who contributed to set up the 1961–1990 sunshine duration database. Strictly in alphabetical order, we acknowledge ARSO-Slovenia, CRA-CMA, Italian Air Force, Meteo France, Meteo Swiss, MIPAF, SCIA-APAT, and ZAMG-Vienna. We thank JRC-GEM for the land cover grids. We are glad to thank all the researchers, collaborators, and volunteers who, over the years, helped us in collecting and quality checking the data: Matteo Cella, Gianluca Lentini, and Veronica Manara. This study has been carried out in the framework of the EU project ECLISE (265240).

Edited by: I. Auer
Reviewed by: two anonymous referees

References

Akinoglu, B. G. and Ecevit, A.: Construction of a quadratic model using modified Angstrom coefficients to estimate global solar radiation, Sol. Energy, 45, 85–92, 1990.

Andretta, A., Bartoli, B., Coluzzi, B., Cuomo, V., Francesca, M., and, Serio, C.: Global solar radiation estimation from relative sunshine hours in Italy, J. Appl. Meteorol., 21, 1377–1384, 1982.

Black, J. N., Bonython, C. W., and Prescott, J. A.: Solar radiation and the duration of sunshine, Q. J. Roy. Meteor. Soc., 80, 231–235, 1954.

Brunetti, M., Lentini, G., Maugeri, M., Nanni, T., Auer, I., Böhm, R., and Schöner, W.: Climate variability and change in the Greater Alpine Region over the last two centuries based on multivariable analysis, Int. J. Climatol., 29, 2197–2225, 2009.

Cat Berro, D., Ratti, M., Mercalli, L., and Lupotto, E.: Il clima di Vercelli dal 1871, Edizioni Mercurio, 2005 (in Italian).

Chung, U. and Yun, J. I.: Solar irradiance-corrected spatial interpolation of hourly temperature in complex terrain, Agr. Forest Meteorol., 126, 129–139, 2004.

Coppolino, S.: A new correlation between clearness index and relative sunshine, Renew. Energ., 4, 417–423, 1994.

Daly, C., Gibson, W. P., Taylor, G. H., Johnson, G. L., and Pasteris, P. A.: A knowledge-based approach to the statistical mapping of climate, Climate Res., 22, 99–113, 2002.

Daly, C.: Guidelines for assessing the suitability of spatial climate data sets, Int. J. Climatol., 26, 707–721, 2006.

Erbs, D. G., Klein, A. A., and Duffie, J. A.: Estimation of the diffuse radiation fraction for hourly, daily and monthly-average global radiation, Sol. Energy, 28, 293–302, 1982.

Gopinathan, K. K. and Soler, A.: Diffuse radiation models and monthly-average, daily, diffuse data for a wide latitude range, Energy, 20, 657–667, 1995.

Gueymard, C. A.: Parameterized transmittance model for direct beam and circumsolar spectral irradiance, Sol. Energy, 71, 325–346, 2001.

Henderson-Sellers, A. and Wilson, M. F.: Surface albedo data for climatic modelling, Rev. Geophys. Space Phys., 21, 1743–1778, 1983.

Hummel, J. R. and Reck, R. A.: A global surface albedo model, J. Appl. Meteorol., 18, 239–253, 1979.

Kasten, F., Golchert, H. J., Dogniaux, R., and Lemoine, M.: European Solar Radiation Atlas (Vol. 1), Verlag Tuv Rheinland, ISBN: 3-88585-195-4, 1984.

Kasten, F. and Young, A. T.: Revised optical air mass tables and approximation formula, Appl. Optics, 28, 4735–4738, 1989.

Iqbal, M.: An introduction to solar radiation, Academic Press, Orlando, FL, OSTI ID: 5596615, 1983.

Landsberg, E. H.: The Urban Climate, Academic Press, Maryland, 1981.

Lavagnini, A. and Jibril, Z.: Monthly maps of daily diffuse solar irradiance for Italy, Renew. Energ., 1, 779–789, 1991.

Newland, F. J.: A study of solar radiation models for the coastal region of South China, Sol. Energy, 43, 227–235, 1989.

Ohmura, A.: Observed long-term variations of solar irradiances at the Earth's surface, Space Sci. Rev., 125, 111–128, doi:10.1007/s11214-006-9050-9, 2006.

Page, J. K.: The estimation of monthly mean values of daily total short-wave radiation on vertical and inclined surfaces from sunshine records for latitudes 40° N–40° S, Proc. UN Conference on New Sources of Energy, 4, 378–390, 1964.

Petrarca, S., Cogliani, E., and Spinelli, F.: La Radiazione Globale al Suolo in Italia. Anni 1998–1999 e media 1994–1999, Edizioni ENEAISBN: 88-8286-055-8, http://clisun.casaccia.enea.it/Pagine/Radiazione.htm, 2000 (in Italian).

Reindl, D. T., Beckman, W. A., and. Duffie, J. A.: Diffuse fraction correlations, Sol. Energy, 45, 1–7, 1990.

Rietveld, M. R.: A new method for estimating the regression coefficients in the formula relating solar radiation to sunshine, Agr. Meteorol., 19, 243–252, 1978.

Rigollier, C., Bauer, O., and Wald, L.: On the clear sky model of the ESRA – European Solar Radiation Atlas – with respect to the HELIOSAT method, Sol. Energy, 68, 33–48, 2000.

Scharmer, K. and Greif, J.: The European Solar Radiation Maps. Vol 1: fundamental and Maps. Le Presses de l'Ecole de Mines, Paris, 2000.

Suri, M. and Hofierka, J.: A new GIS-based solar radiation model and its application to photovoltaic assessments, Transactions in GIS 8, 175–190, 2004.

Spinoni, J.: 1961–90 High-Resolution temperature, precipitation, and solar radiation climatologies for Italy, Ph.D. thesis, Milan University, available at: http://air.unimi.it/bitstream/2434/155260/2/phd_unimi_R07883_1.pdf, 2010.

USGS: GTOPO30 Digital Elevation Model, http://eros.usgs.gov/#/Find_Data/Products_and_Data_Available/gtopo30_info, 1996.

Wild, M.: Global dimming and brightening: a review, J. Geophys. Res, 114, D00D16, doi:10.1029/2008JD011470, 2009.

Wild, M., Gilgen, H., Roesch, A., Ohmura, A., Long, C. N., Dutton, E. G., Forgan, B., Kallis, A., Russak, V., and Tsvetkov, A.: From dimming to brightening: decadal changes in solar radiation at Earth's surface, Science, 308, 847–850, 2005.

Wilson, M. F. and Henderson-Sellers, A.: A global archive of land cover and soils data for use in general circulation models, Int. J. Climatol., 5, 119–143, 1985.

10

A monthly precipitation database for Spain (1851–2008): reconstruction, homogeneity and trends

M. Y. Luna, J. A. Guijarro, and J. A. López

Agencia Estatal de Meteorología, Leonardo Prieto Castro, 8, 28040 Madrid, Spain

Correspondence to: M. Y. Luna (mlunar@aemet.es)

Abstract. The compilation and reconstruction of a dataset integrated by 66 long monthly precipitation series, covering mainland Spain and the Balearic Islands, is presented. The reconstruction is based on the hypothesis that the cessation of data recording at one observatory is followed by the establishment of a new observatory very close to the closed one. In order to detect and adjust for possible multiple change points or shifts that could exist in the precipitation series, the R-package CLIMATOL V2.0 is used. This method enables to take advantage of the whole historical Spanish precipitation network in the detection and correction of inhomogeneities. The analysis of annual precipitation trends indicate a high temporal variability. Negative trends dominate for the period 1951–2008 but not for all observatories. On the other hand, positive trends can be detected in the northern Spain for 1902–2008.

1 Introduction

Global warming will likely to cause changes in the hydrological cycle and, hence, in the precipitation. Even if the total amount of precipitation does not change, rainy days frequency and precipitation rates will likely to change (IPCC, 2007). According to model predictions, the precipitation in the Mediterranean region will decrease in response to anthropogenic global warming because of the northward migration of the Hadley cell. In addition, the global climate simulations predict a precipitation reduction for the 21st century. However, a general decrease in precipitation has not been detected in Spain during the 20th century: the large natural interannual variability and the short length of instrumental records impede the detection of possible trends (CLIVAR España, 2011).

There is a clear necessity of a set of long time precipitation series for Spain that extend over the whole 20th century. A statistically robust analysis needs historical precipitation series that can include temporal periods with negligible anthropogenic forcing. The purpose of this study is to extend, revise, improve and update the Spanish monthly/annual centenarian precipitation series in an earlier work at the Spanish Meteorological Agency by Almarza et al. (1996). Historical precipitation series is aimed to obtain in order to achieve

more consistent variability and trend analyses. In this paper, a dataset integrated by 66 long monthly precipitation series covering mainland Spain and the Balearic Islands is compiled to allow the research community to use it as a reference for climatic studies, and a preliminary analysis of their trends is performed.

2 Data and reconstruction process

The data used have been extracted from the historical data base of the Spanish Meteorological Agency (AEMET: **A**gencia **E**statal de **MET**erología). The precipitation series stored in the historical data base of AEMET are very irregular, differing in length, missing data amount, beginning and end dates, etc, and therefore they are difficult of analyze in its original form. Nevertheless, they contain enough information to have the possibility of creating a reconstructed database. A single long time series is constructed from a number of shorter series belonging to nearby observatories. The reconstruction is based on the hypothesis that the cessation of data recording at one observatory is followed by the establishment of a new observatory very close to the closed one. If the observatories are very close, the differences in monthly precipitation amounts are usually very small. The reconstructed series is attributed to the last observatory that

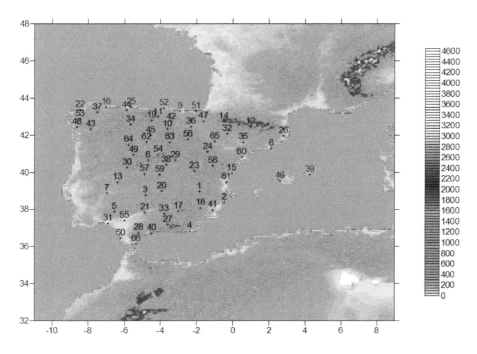

Figure 1. Geographical distribution of the 66 precipitation series in Spain. The orography is displayed.

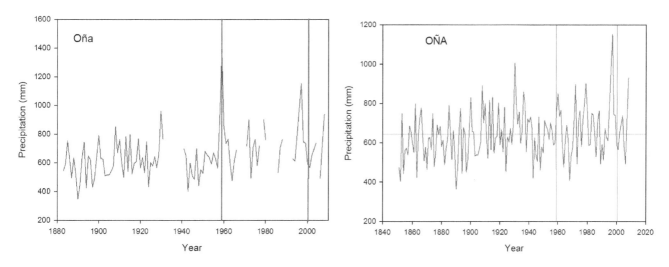

Figure 2. Example of the homogenisation and reconstruction process for Oña observatory (no 42 in Fig. 1). (**a**) Recorded data and the breaks detected in 1959 and 2000. (**b**) Reconstructed time series.

is nowadays working and will be probably working in the future. In this way, 65 series have been reconstructed distributed around the mainland Spain and the Balearic Islands. These time series have records greater than 90 yr and the number of missing values is less than 10 %, most of them during the Spanish civil war. A metadata archive that contains the information about relocations and other technical changes is generated. Figure 1 displays the spatial distribution of the precipitation series and the topography of the studied area, covering peninsular Spain and the Balearic Islands. In this figure, it can be observed that Spain is located

between two very different water bodies, the Atlantic Ocean and the Mediterranean Sea, and the mountain ranges are oriented mainly W-E in the western peninsular area and N-S near the eastern coast, making the spatial precipitation patterns highly variable. In this geographical environment, three pluviometric areas have been identified: one in the north-eastern part, another covering the south-western and central area and finally, the area covering the Mediterranean coast and the south-eastern (Morata el al., 2006).

Table 1. Inhomogeneities detected by CLIMATOL package.

NAME	NO BREAKS	DATE 1	DATE 2	DATE 3
OÑA-IBERDUERO	3	2000	1959	1958
PAMPLONA OBSERVATORIO	2	1941	1933	
CIUDAD REAL (ESCUELA DE MAGISTERIO)	2	1880	1872	
GRANADA (BASE AEREA)	2	1939	1938	
BUSTAMANTE PANTANO	2	1994	1989	
LEON (VIRGEN DEL CAMINO)	2	1920	1906	
SAN SEBASTIAN (IGUELDO)	2	1921	1925	
A CORUÑA (ESTACION COMPLETA)	2	1942	1935	
CACERES (CARRETERA TRUJILLO)	2	1914	1910	
OURENSE (GRANXA DEPUTACION)	1	1969		
LINARES DEL ARROYO	1	1962		
AVILA (OBSERVATORIO)	1	1907		
LUGO (COLEXIO FINGOI)	1	1985		
SEGOVIA (OBSERVATORIO)	1	1919		
SEVILLA (AEROPUERTO)	1	1875		
GERRI DE LA SAL	1	1995		
HUELVA (RONDA ESTE)	1	1936		
SALAMANCA (MATACAN)	1	1911		
SANTIAGO COMPOSTELA (LABACOLLA)	1	1929		
ALICANTE (CIUDAD JARDIN)	1	1857		

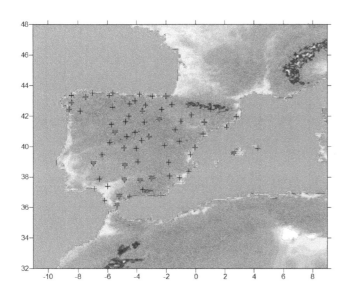

Figure 3. Significative trends for the 1902–2008 period: red triangles (blue crosses) indicate negative (positive) trends.

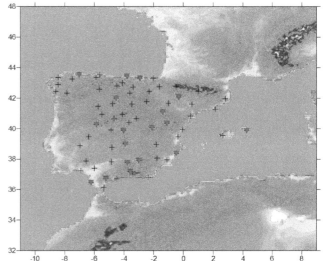

Figure 4. Significative trends for the 1951–2008 period: red triangles (blue crosses) indicate negative (positive) trends.

Data from Gibraltar observatory are included in order to use the information contained in a series that began in the 19th century. The Gibraltar data were obtained from The Global Historical Climatology Network (GHCN-Monthly) database of the Climate Services and Monitoring Division (NOAA).

3 Homogenisation process

Logically, the resulting combined series can exhibit inhomogeneities which must be identified and removed before any further analyses. In order to detect and adjust for possible multiple change points or shifts that could exist in the precipitation series, the R-package CLIMATOL 2.0 (Guijarro, 2011) has been used. This method enables to take advantage of the whole Spanish precipitation network in the detection and correction of inhomogeneities. The homogeneity tests are applied on a difference series between the problem station and a reference series constructed as a weighted average of series from nearby stations. The selection of these stations is only based on proximity in order to use the nearest stations even when no or too short common period of observations compromise the computation of correlation coefficients. This homogenisation methodology involves three steeps: a type II regression (Sokal and Rohlf, 1969), missing data estimation and outlier and break detection and correction (Guijarro, 2011).

The results of the homogenisation process are displayed in Table 1. Only 20 of 66, the 30 %, series have been found inhomogeneous and 9 of them present two breaks. In Oña observatory (in the north of the peninsular area) CLIMATOL detects two very close breaks, maybe due to the missing data. The time series for Oña is displayed in Fig. 2 before and after the reconstruction process as an example.

4 Trend analysis results

The statistical significance of the trends has been obtained for all observatories by the Mann-Kendall statistical test for two different periods: 1902–2008 and 1951–2008. The first temporal period encompasses the whole 20th century and the second one is elected in order to compare the results with similar analysis from different authors. In general, it can be seen in Figs. 3 and 4 that the trend is significant in some sites, but not everywhere. For the 20th century (Fig. 3), there are significant positive trends in the north of the peninsular area and negative ones in the south and the Balearics. These results are in agreement with precipitation variability studies compiled and described in Cuadrat and Martín-Vide (2007). The negative trend detected in Gibraltar, one of the longest records, has been detected by other authors (CLIVAR España, 2011). However, for the second half of the 20th century (Fig. 4), the observed significant trends are negative, except for Avila observatory. This observatory is situated at 1130 m of altitude in the north slope of a mountain chain (Sistema Central) and the positive value could be due to orographic effects. Only 17 of the 66 observatories show significant negative trends. In both time periods, the spatial distribution of the negative or positive trends does not correspond to any Spanish pluviometric area described in Morata el al. (2006).

5 Conclusions

A historical monthly precipitation data base has been constructed by means of CLIMATOL package that provides a useful tool for homogenization and missing value estimation taking into account the whole climatic information available in the historical Spanish climatic database. There is no overall valid significant trend in the historical precipitation series in Spain: it depends on the analysed period.

There is a clear negative trend for the period 1951–2008 consistent with other authors. The 60's and 70's decades were exceptionally rainy and the following dry decades contributed to this negative trend that is not reproducible if centenarian series are analyzed. The historical precipitation dataset created in this work does not exhibit a significant reduction in precipitation amounts. The anthropogenic precipitation signal in this region predicted by climate models (CLIVAR España, 2011) is not detectable in the analysis of this instrumental database.

Acknowledgements. The authors thank the Climate Services and Monitoring Division (NOAA) for Gibraltar observatory data (http://www.ncdc.noaa.gov/ghcnm/).

Edited by: M. Brunet-India
Reviewed by: two anonymous referees

References

Almarza, C., López-Díaz, J. A., and Flores, C.: Homogeneidad y variabilidad de los registros históricos de precipitación de España, Monografía Técnica A-143 Instituto Nacional de Meteorología, Ministerio de Medio Ambiente, ISBN: 8449802210, 1996.

CLIVAR España: Climate in Spain: past, present and future, Regional climate change assessment report, edited by: Pérez, F. F. and Boscolo, R., Spain, 85 pp., ISBN: 9788461481156, http://www.clivar.es, 2011.

Cuadrat, J. M. and Martín-Vide, J. (Eds.): Spanish Climatology. Past, Present and Future, P.U.Z. Zaragoza, 575 pp., 2007.

Guijarro, J. A.: User's guide to Climatol V2.0. An R contributed package for homogenization of climatological series (and functions for drawing wind-rose and Walter&Lieth diagrams), http://webs.ono.com/climatol/climatol.html, 2011.

IPCC: Climate Change 2007: The Physical Science Basis. Contribution of Working Group I to the Fourth Assessment Report of the Intergovernmental Panel on Climate Change, Cambridge University Press, Cambridge, United Kingdom and New York, NY, USA, 996 pp., 2007.

Morata, A., Martín, M. L., Luna, M. Y., and Valero, F.: Self-Similarity patterns of precipitation in the Iberian Peninsula, Theor. Appl. Climatol., 85, 41–59, 2006.

Sokal, R. R. and Rohlf, F. J.: Introduction to Biostatistics, 2nd edn., edited by: Freeman, W. H., New York, 363 pp., 1969.

11

A study of the 1 and 2 January 2010 sea-storm in the Ligurian Sea

F. Pasi[1,2], A. Orlandi[1,2], L. F. Onorato[3], and S. Gallino[3]

[1]CNR-IBIMET, Firenze, Italy
[2]Consorzio LaMMa, Sesto F. no, Italy
[3]Meteo Hydrological Center, Genova, Italy

Abstract. During the last days of 2009 and the first days of 2010, a wide and deep low pressure system over Western Europe generated a very extended and strong southerly pressure gradient on the whole Western Mediterranean Sea with a resulting very rough to high sea state. Over the Ligurian Sea (North Western Mediterranean) the resulting sea state was a combination of a very tuned (in both frequency and direction) swell coming from the south-west, with nearly oceanic peak wave period, and a broader north-westerly wind sea with shorter period. This kind of sea state, not extreme in terms of significant wave height, caused unusual widespread damages to Ligurian coastal structures.

In this study, authors investigated the structure of such a combined sea state by analysing numerical weather prediction outputs coming from atmospheric and wave models and comparing them with data coming from ondametric buoys and meteorological stations located in the Ligurian Sea area. As a result, it was found that the forecasting model chain almost correctly predicted the wave height in a first phase, when the sea state was only due to the first south-westerly swell peak, while significantly underestimated the combined sea state, when also the second north-westerly wind sea developed and interacted with the first one.

By analysing the structure of directional wave spectra forecasted by the operational wave model and measured by the buoys, authors have attempted to find out the reasons for model deficiencies in forecasting the time evolution of significant parameters characterising the sea state.

1 Introduction

Sea storms in the Mediterranean basin are a quite common event especially during winter months (Littmann, 2000). In particular, over the Lion Gulf (due to frequent Mistral outbreaks) and the Genoa Gulf (one of the most cyclogenetic areas of the Mediterranean, Trigo et al., 2002), several episodes of significant wave height (hereafter SWH) above 3 m (Medatlas, 2004) are recorded each year. A climatological study performed at the Regional Weather Service of Liguria Region (Onorato et al., 2006 and internal report), analysing the data recorded by the Ventimiglia buoy (here-after XXM, Fig. 1) during the period 1998–2010, found that sea-storms with SWH greater than 4.3 m are correlated with a mean wave period (hereafter MWP) of about 8.4 s.

The present study is focused on an interesting sea-storm that occurred in the first days of 2010 with an unusual damage producing capability over the Ligurian coasts. Damages were widespread and affecting a wide portion of the Ligurian coast, comprising the most western tip, very rarely affected by such events (see red line in Fig. 1). The Ligurian Regional Government reported the interruption of railway and "Aurelia" main road and the damaging of various harbor docks and several commercial activities, for a total of more than 20 millions of euro. The Ligurian Regional Weather Service on 1 January 2010 morning issued a level 1 warning (the lowest of 3 levels), which "a posteriori" must be considered underestimated.

Figure 1. Detailed map of the Ligurian sea and coastlines, where red line describes the area of main coastal damages. Blue dots show the location of the two moored buoys used in the study, the Ventimiglia (XXM) and Tuscany Region one (RT).

Trough a detailed analysis of the evolution of the sea state occurred during the event, authors will try to answer the following questions: (1) among the swell characteristics which one was mainly responsible for such a severe impact over the Ligurian coast? (2) had the forecasters enough information to make a "good job" or did they need something else to estimate the risky forthcoming events?

The present work is organized as follows: Sect. 2 contains a detailed description of the sea-storm with the support of several data collected by ondametric buoys. Section 3, the most articulated one, contains: a brief description of the operational modeling chain in use by the two Regional Meteorological Weather Services of Liguria and Tuscany (in charge of issuing warnings), a discussion of the models' performances in terms of average fields and a general discussion of the sea-storm dynamic trough a spectral analysis approach. Finally, in Sect. 4, some conclusions and suggestions for the forecasters' operational practice are drawn.

2 Description of the event

On 29 December 2009 a deep (around 980 hPa) low pressure system, initially located in the Atlantic off the Irish coasts, begun to extend its influence to the Western Mediterranean Sea with a pressure thalweg. During 30 and 31 December 2009 the low pressure system moved south-eastward across France towards the Lion Gulf, filling up to 990 hPa: an extended gale-force south-westerly pressure gradient was present over the majority of Western Mediterranean basin.

On 1 January 2010 the low pressure moved quite rapidly from the Lion Gulf (990 hPa at 00:00 UTC) to the Genoa Gulf (988 hPa at 06:00 UTC), then passing to the Adriatic Sea (992 hPa at 18:00 UTC) and finally moving over the Balkans (990 hPa at 00:00 UTC on 2 January 2010). As a consequence the south-westerly pressure gradient in the eastern part of Western Mediterranean Sea (see Fig. 2a) experienced first an intensification, then a westerly to north-westerly veering and finally a rapid decrease on 2 January 2010 second half of the day (see Fig. 2b).

An interesting feature of the wind field is that, while in the western part of the Mediterranean Sea south-westerly and westerly gale force winds persisted for more than 48 h, in the area of Genoa Gulf light winds dominated. The XXM buoy, located in open sea (see Fig. 1), recorded fluctuating gentle to moderate south-westerly winds (10–17 kt) until 00:00 UTC on 2 January 2010, increasing to strong or near gale winds (25–35 kt) only after the west, north-west veering and persisting until 15:00 UTC on 2 January 2010 (see Fig. 3). Also coastal wind gauge stations confirmed the presence of a light wind regime, with sustained wind speed values only after the rotation to north on 2 January 2010 (not shown).

For what concerns the sea-storm characteristics, data from the XXM buoy allow the individuation of three different phases (see Fig. 3). The first phase that goes from 12:00 UTC on 31 December 2009 to 18:00 UTC on 1 January 2010 was characterised by an increase in both SWH and MWP (from 2 to 4 m and from 6 to 11 s respectively). The corresponding sea state can be ascribed to the arrival of a swell coming from a very distant generating area, probably located in Western Mediterranean Sea (south of Balearic Islands). In fact, a wave group with a period of about 10 s propagates at a speed of about 25 kt and covers the distance between the generating area and the XXM buoy (about 400 nautical miles) in 16 h. Considering that the XXM buoy recorded the first peak in SWH values at 15:00 UTC on 1 January 2010, the generation time of this peak should be around 00:00 UTC on 1 January 2010. This seems in good correlation with models' outputs forecasting a 10 s period, 6 m SWH sea state in generation south of Balearic Islands (not shown).

The second phase, which goes from 18:00 UTC on 1 January to 06:00 UTC on 2 January 2010, was characterised by SWH and MWP remaining almost constant at values of 4 m and 11 s respectively. The third phase, which goes from 06:00 UTC to 18:00 UTC on 2 January 2010, was characterized by an irregular decrease of SWH and a more regular decrease of MWP. Each of these sea-storm's phase show its own peculiarities and will be discussed more in detail in the next paragraphs.

For what concerns coastal damages the most effective of the three phases was the second one (between 18:00 UTC on 1 January to 06:00 UTC on 2 January 2010).

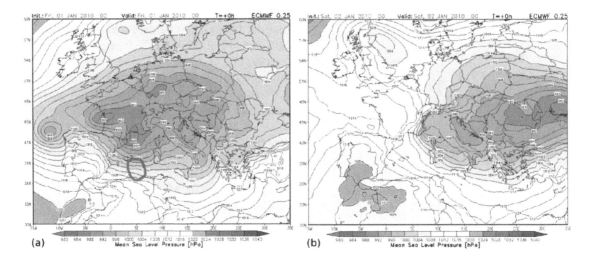

Figure 2. Mean Sea Level Pressure (hPa) from ECMWF analysis (bluish colors for low pressure values). Left panel **(a)** 00:00 UTC on 1 January 2010 with main low pressure system over France and Lion Gulf at 990 hPa, strong westerly baric gradient in the Central Mediterranean Sea. Red Circle indicates supposed generation area of swell component arrived in Ligurian Sea on 1 January 2010 early afternoon. Right panel **(b)** 00:00 UTC on 2 January 2010 with main low pressure system over Balkans at 990 hPa and strong north-westerly baric gradient over Central Mediterranean Sea.

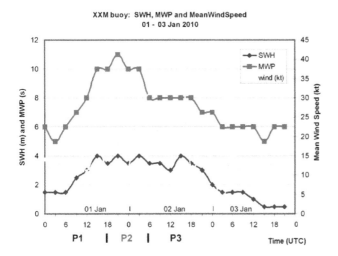

Figure 3. 1–3 January 2010 (hours in UTC) XXM buoy time-serie. Blue line indicates SWH (m), purple line MWP (s), yellow line 10 m wind speed (kt). P1 to P3 refers to the sea-storm different phases. P2 (in red) is the most damaging one.

Only considering SWH values, the total amount of damages reported (see Sect. 1) is quite inexplicable because: (1) sea-storms with SWH of around 4 m do happen several times a year (about 30 cases were observed at XXM buoy in the last 12 yr); (2) sea-storm phase 3 was not damaging despite SWH was still around 4 m (as in phase 2); (3) the wind regime in the area was light and surely not recognised as a dangerous one. Nevertheless, if the 12 yr statistics of MWP occurrence at XXM buoy are analysed, the actual 11 s value is the largest recorded.

Another consideration, which will not be discussed in detail in this article, but that is surely worth to mention was a coincident high astronomic tide that clashed with the strong barometric high tide (mean sea level pressure was quite low) causing record sea level rise up to 60 cm during the sea-storm peak (Ray et al., 2011; and Padman and Erofeeva, 2004). This contributed to coastal wave penetration and might also explain why local reporters visually estimated the wave height, when breaking at coast at more than 6 m (Wang et al., 2008).

3   Modeling results

4   The operational chain

The modeling chain applied here to investigate the sea-storm details is the one implemented and used operationally at the Regional Weather Service of Tuscany region, namely the Laboratory for Meteorology and environmental Modeling (Consorzio LaMMA, http://www.lamma.rete.toscana.it) and is composed by a meteorological model and a wave model running in cascade. The former is the atmospheric model WRF-NMM (http://www.wrf-model.org, Janiic, 2003 and Skamarock et al., 2005), at a resolution of 0.12° with initial and boundary conditions from NCEP-GFS (T382L64) at 0.5° resolution. The latter is the WW3 third generation spectral wave model (http://polar.ncep.noaa.gov/waves/wavewatch/wavewatch.shtml, Tolman et al., 2002; Tolman, 2006, 2009; Komen et al., 1994; Janssen, 2007), running over the whole Mediterranean Sea at the same resolution of the atmospheric model. Atmospheric initial and boundary conditions for the present study was taken from the

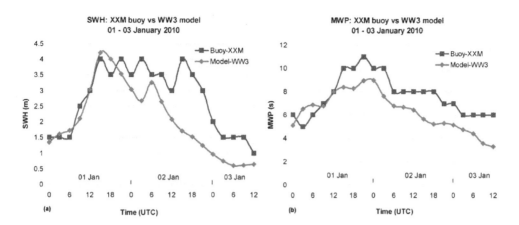

Figure 4. Comparison between XXM buoy (blue line) and ww3 model (orange). Time series for 1–3 January 2010 (hours in UTC). Left panel **(a)** SWH (m), right panel **(b)** MWP (s).

1 January 2010, 00:00 UTC operational GFS run, as a consequence the maximum sea-storm peak (18:00 UTC on 1 January to 06:00 UTC on 2 January 2010) is at the +18 to +30 h of model forecast.

4.1 Average fields

Average quantities like SWH and MWP are the most commonly used by forecasters in the operational practice when elaborating sea state bulletins and emergency warnings for navigation (Niclasen et al., 2011) and coastal structures (US Army CEM, http://chl.erdc.usace.army.mil/cem). Figure 4a and b show a comparison between model outputs (SWH and MWP) and corresponding observed data at XXM buoy. For sea-storm phase 1 (until 18:00 UTC on 1 January 2010), the wave model was quite good in forecasting SWH (matching both time phase and intensity), but underestimated MWP (forecasted at 8 s against the measured 10 s). In terms of the sea-storm character, limiting the analysis to such model data, the forecaster could be brought to think to a sea state more dominated by wind sea, or a sea state due to a swell being weaker or coming from a nearer place than it has been in reality. For sea-storm phase 2 (18:00 UTC on 1 January to 06:00 UTC on 2 January 2010) the wave model underestimated both SWH and MWP, even though MWP is forecasted slightly growing. This last feature is compatible with a sea-storm that is getting more and more swell-dominated. Model predicted MWP reaches its maximum between 21:00 UTC on 1 January and 00:00 UTC on 2 January 2010, with corresponding SWH of 3.5 and 3 m, against measured values of 4 m. Again the wave model is probably underestimating the intensity of the swell component of the sea state, coming from the distant generation area (Balearic Islands). This is probably due to a not perfect representation of the wind field over the sea-storm fetch. This, in turn, might depend either on a general underestimation of the wind speed or to a not correct wind direction representation (e.g. a more

westerly than south-westerly component) or both. These aspects should be investigated more deeply with a more general (in the sense of the area involved) verification against observed data. Sea-storm phase 3 is the less accurate predicted one, with a strong underestimation of SWH and MWP. In this phase winds veered from the northern direction (North-West, North) and increased to gale-force (as confirmed by a second observed SWH peak of 4 m at around 15:00 UTC on 2 January 2010). This rapid increase is not predicted by the model and should be investigated more deeply.

4.2 Analysis of wave directional spectra

The analysis of the directional wave spectrum allows the most complete investigation of predicted sea states (Orlandi et al., 2008; Bradbury et al., 2007). Figure 5a and b show model energy spectra of the predicted sea state during phase 2 and phase 3 of the sea-storm at XXM buoy.

In this representation, the color palette is indicative of the wave energy density ($m^2\,s\,rad^{-1}$) and the meteorological convention is used to

specify wind direction as that from which the wind blows. On the north-west (north-east) radius the period (the frequency) is reported with a graduation that increases (decreases) moving away from the plot center. On the south-west radius the wave length graduation (in meters) is reported, while on the south-east radius the group velocity graduation (in knots) is shown. Super imposed blue arrow is the local mean wind direction, green arrow is the mean wave direction and red arrow is the wave peak direction.

The possible sea state scenario that can be reconstructed on the basis of such model data, is described below.

Figure 5a valid for 00:00 UTC on 2 January 2010, corresponds to the sea-storm phase 2 (widespread coastal damages). This is an unimodal spectrum, which is a sea state with a single south-westerly component. The main features of this sea state are: (1) a light wind (blue arrow) coming from

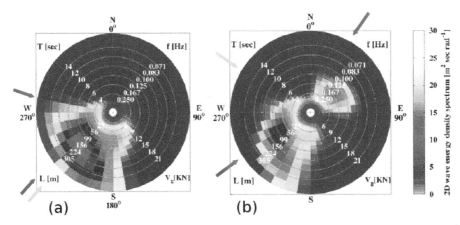

Figure 5. Wave energy spectrum for ww3 model at XXM buoy location. Color is proportional to wave energy ($m^2 s rad^{-1}$). Blue arrow indicates the mean wind direction, green arrow the mean wave direction and red arrow the wave peak direction. Left panel **(a)** valid for 00:00 UTC on 2 January 2010 (sea-storm phase 2). Right panel **(b)** valid for 15:00 UTC on 2 January 2010 (sea-storm phase 3).

the west, north-west (not generating a relevant wave component); (2) nearly coincident mean and peak wave direction (from the south-west, green and red arrows); (3) mean and peak angular spreading coincident and very narrow (around 12°); (4) a very intense and tuned peak at about 11 s of period. Looking carefully, the structure of the peak is asymmetric and seems to be composed of two sub components: one weaker satellite component, more southerly and with a period of near 12 s (probably coming from a farther more intense generation area), and the other dominant and narrower in frequency component, from south-west, with a slightly smaller period (near 11 s). These two peaks components might correspond to two different phases of the low pressure development and or movement, with maxima winds producing two different distant wave generation areas between Balearic and Sardinia Islands. All these features indicate that the predicted sea state (even though slightly underestimated) consisted of a very tuned (in both direction and frequency) and powerful swell coming from the south-west with nearly oceanic period (not usual in the Mediterranean Sea and especially in the Genoa Gulf), totally uncorrelated with the local weak wind (blue arrow), which is no relevant wind sea components are present.

Figure 5b refers to 15:00 UTC on 2 January 2010, which is sea-storm phase 3 (not causing coastal damages), when a second SWH maximum was registered (about 4 m). This is a 3 peaks energy spectrum with the south-westerly component still being the most energetic, but now lower than in phase 1 and with lower peak period (about 8 s).

Main features of this sea state are: (1) a short period (around 6 s) not very energetic sea state from the north-east caused by a moderate to strong north-easterly wind (blue arrow); (2) a persistent, but lower, main peak (red arrow) from the south-west with little (but increasing) angular spreading; (3) a still present satellite component from the south, southwest, probably older (longer period) and less energetic that

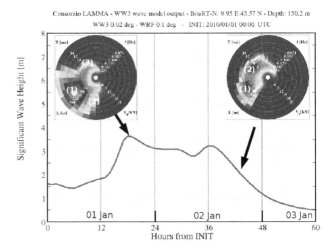

Figure 6. SWH (m) wave model forecast time-serie for the RT buoy location valid for the period from 00:00 UTC on 1 January to 12:00 UTC on 3 January 2010. Upper left: directional wave spectrum forecast for 18:00 UTC on 1 January as indicated by the black arrow. Upper right: same as upper left, but for 12:00 UTC on 2 January 2010.

the main one; (4) a very wide total mean angular spreading (90°) resulting in a total mean wave direction from the north-west (green arrow, not realistic being the average of the whole spectrum).

These considerations indicate that sea-storm phase 3 mean SWH was the combination of at least two nearly opposite sea states (one from the south-west the other from the northeast), and that the main south-westerly peak of the sea-storm was less energetic, less tuned and, more relevant, with lower peak period than in phase 2. These are probably among the characteristics that might explain why sea-storm in phase 3 was less effective in penetrating and damaging the coast than in phase 2.

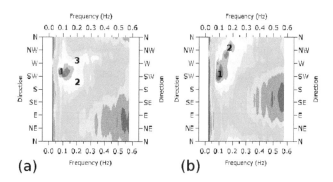

Figure 7. RT buoy directional wave energy spectrum. Colour is proportional to wave energy ($m^2\,s\,rad^{-1}$). Super imposed numbers (black) indicate the main spectral peaks (see text). Left panel **(a)** 12:00 UTC on 1 January 2010 (arrival of the distant swell). Right panel **(b)** 19:00 UTC on 2 January 2010 (decaying phase). The graphical format of these plots is different from the model's directional plots because buoy data are processed by a proprietary software of Datawell (Waves@21) that allows only this kind of representation.

A further analysis has been performed by comparing model spectra with those measured by a Datawell MKIII directional wave buoy installed by Tuscany Region north of Gorgona Island (RT-buoy, see Fig. 1). In Fig. 6 model's forecasted time-series of SWH together with two model wave spectra are shown. The first spectrum is valid for 18:00 UTC on 1 January 2010, i.e. the time of the first peak due to the arrival of the distant swell. A comparison with the corresponding spectrum recorded by the RT-buoy (Fig. 7a) show a good agreement of the main features: a dominant peak (1) from the south-west (swell), a minor component from the south (2) and an even minor component from the north-west (3), which is correlated with the local wind direction. Also the values of the period of the 3 peaks are in fair agreement while SWH values are slightly overestimated (model 3.5 m, buoy 2.9 m).

The second spectrum is valid for 19:00 UTC on 2 January 2010, i.e. during the decaying phase. A comparison with the corresponding spectrum recorded by the buoy (Fig. 7b) shows in both spectra two peaks in general agreement: a swell peak from south-west (1), a peak from northwest correlated with the wind (2). The model underestimate the peak from the south-west and the corresponding value of SWH is significatively lower than the recorded one (model 2.1 m, buoy 3.7 m).

5 Conclusions

In this work, the authors investigated a sea-storm that occurred during the first days of year 2010 in the Western Mediterranean Sea. The peculiarity of this event was that a not unusual sea-storm (more than 3 times per year for similar SWH values) caused unusual widespread damages to a

wide portion of the Ligurian coast. With data collected at two ondametric buoys and ground stations (not shown) and with numerical outputs from an operational met-ocean modelling chain (WRF+WW3), the sea-storm sea state structure has been investigated. As a result, it was found that the sea-storm evolution could be divided into three different phases. Only during phase 2 relevant coastal damages was produced. The sea state of this sea-storm phase was characterised by an extreme (relatively to the Mediterranean wave climate) peak wave period of 11 s (the only occurrence in the last 12 yr).

In the second part of the study, operational forecasts were used to investigate more deeply the sea-storm characteristics. The pro and cons of assuming an average versus a full spectral approach is briefly described in both phase 2 and 3 of the sea-storm. It comes out that the full spectral approach is by far more complete allowing a better understanding of the sea state structure. As a consequence, it is shown why phase 3 of the sea-storm, even though characterized by the same SWH values of phase 2, was not as effective in producing coastal damages. Finally it is suggested the adoption of a full spectral approach also in the operational met-ocean forecasting practice.

From the point of view of models' performance, it is necessary to further investigate and improve the operational models chain. Even though correctly representing the overall evolution of a very rough to high sea state in the open sea over Western Mediterranean Sea, the chain was not able to correctly extend the intensity of the swell to the target area (both for SWH and MWP).

The influence of local bathymetry might have also played an important role in the waves' effectiveness to penetrate the coast (local observer reported 5 to 6 m height waves). This should be addressed with specific high resolution modeling (e.g., SWAN, Booij et al., 1999), which is under development at Consorzio LaMMA. During this further investigation the possible influence of coupled astronomic and barometric tides at coast should be taken into account, as well.

Acknowledgements. This work is part of a long-term cooperation between the Meteo-hydrological service of Liguria Region (www.meteoliguria.it) and the Regional Weather Service of Tuscany region (www.lamma.rete.toscana.it). We acknowledge Meteo France for the XXM buoy data.

Edited by: F. Stel
Reviewed by: M. Sioutas and another anonymous referee

References

Booij, N., Ris, R. C., and Holthuijsen, L. H.: A third-generation wave model for coastal regions. Model description and validation, J. Geophys. Res., 104(C4), 7649–7666, 1999.

Bradbury, A. P., Mason, T. E., and Poate, T.: Implications of the spectral shape of wave conditions for engineering design and coastal hazard assessment – evidence from the english channel, 10th International Workshop on Wave Hindcasting and Forecasting and Coastal Hazard Symposium North Shore, abstract no. J5, Oahu, Hawaii, 11–16 November 2007.

Janjic, Z. I.: A Nonhydrostatic Model Based on a New Approach, Meteorol. Atmos. Phys., 82, 271–285, 2003.

Janssen, P. A. E. M.: Progress in ocean wave forecasting, J. Comp. Sci., 227, 3565–3572, 2007.

Komen, G. J., Cavalieri, L., Donelan, M., Hasselmann, K., Hasselmann, S., and Janssen, P. A. E. M.: Dynamics and modelling of ocean waves, Cambridge Book online, Online ISBN: 9780511628955, Hardback ISBN: 9780521470476 , Paperback ISBN: 9780521577816, Book doi:10.1017/CBO9780511628955, 1994 (Online Publication Date: January 2010).

Littmann T.: An empirical classification of weather types in the Mediterranean basin and their interrelation with rainfall, Theor. Appl. Climatol., 66, 161–171, 2000.

Medatlas Group: Wind and Wave Atlas of the Mediterranean Sea, W.E.A.O. Research Cell, 2004.

Niclasen, B. A., Simonsen, K., and Magnusson, A. K.: Wave forecasts and small-vessel safety: A review of operational warning parameters, Mar. Struct., in press, 2011.

Onorato, L., Gemelli, P., and Gallino, S.: A supervisionate classification approach of south-westerly wind regime causing severe weather over the Gulf of Genoa, 6th Annual Meeting of the EMS/6th ECAC, P0095 POSTER EMS2006-A-00386, 2006.

Orlandi, A., Pasi, F., Onorato, L. F., and Gallino, S.: An observational and numerical case study of a flash sea storm over the Gulf of Genoa, Adv. Sci. Res., 2, 107–112, doi:10.5194/asr-2-107-2008, 2008.

Padman, L. and Erofeeva, S.: A Barotropic Inverse Tidal Model for the Arctic Ocean, Geophys. Res. Lett., 31, L02303, doi:10.1029/2003GL019003, 2004.

Ray, R. D., Egbert, G. D. and Erofeeva, S. Y.: Tide predictions in shelf and coastal waters: status and prospects, in: Coastal Altimetry, edited by: Vignudelli, S., Kostianoy, A. G., Cipollini, P., and Benveniste, J., 1st Edn., XII, 566 pp., ISBN 978-3-642-12795-3, Springer-Verlag, 2011.

Skamarock, W. C., Klemp, J. B., Dudhia, J., Gill, D. O., Barker, D. M., Wang, W., and Powers, J. G.. A description of the Advanced Research WRF Version 2, NCAR Tech Notes-468+STR, 2005.

Tolman, H. L.: A third-generation model for wind waves on slowly varying, unsteady and inhomogeneous depths and currents, J. Phys. Ocean., 21, 782–797, 2006.

Tolman, H. L.: User manual and system documentation of wavewatch III version 3.14, NOAA/NWS/NCEP/MMAB Technical Note 276, 2009.

Tolman, H. L., Balasubramaniyan, B., Burroughs, L. D., Chalikov, D. V., Chao, Y. Y., Chen, H. S., and Gerald, V. M.: Development and implementation of wind generated ocean surface wave models at NCEP, Wea. Forecast., 17, 311–333, 2002.

Trigo, I. F., Bigg, G. R., and Davies, T. D.: Climatology of Cyclogenesis Mechanisms in the Mediterranean, Mon. Weather Rev, 130, 549–569, 2002.

Wang, S., McGrath, R., Hanafin, J., Lynch, P., Semmler, T., and Nolan, P.: The impact of climate change on storm surges over Irish waters, Ocean Model., 25, 83–94, 2008.

12

The archive and library of the former Italian Central Office for Meteorology and Climatology

M. C. Beltrano[1], S. Esposito[1], and L. Iafrate[2]

[1]Unità di Ricerca per la Climatologia e la Meteorologia applicate all'Agricoltura, Consiglio per la Ricerca e la Sperimentazione in Agricoltura, Roma, Italy
[2]Centro di Ricerca per lo Studio delle Relazioni tra Pianta e Suolo, Consiglio per la Ricerca e la Sperimentazione in Agricoltura, Roma, Italy

Correspondence to: M. C. Beltrano (mariacarmen.beltrano@entecra.it)

Abstract. The Paper Archive and Library of CRA-CMA (Agricultural Research Council – Research Unit for Climatology and Meteorology applied to Agriculture) are an important source of meteorological data for climate research.

CRA-CMA's Paper Archive gathers a collection of about 850 historical meteorological datasets. Among them, 40 are nowadays still in progress, 260 are more than thirty years long and 20 exceed one century. Moreover, the specialized Library of CRA-CMA gathers several publications containing meteorological data from many Italian and foreign observatories and an important collection of scientific journals and historical books on Atmospheric Sciences, Geophysics and Agrometeorology published both in Italy and abroad and dating from the second half of the sixteenth century. Even if input data for climate models are generally based on 30 yr long datasets, nevertheless longer observational series (up to 50 or 100 yr) are a key element to better understand the climate system behavior.

Until today, the library described in this paper is a CRA-CMA heritage almost unknown to the international scientific community.

1 Introduction

The origin of the current Archive and Library was the establishment in 1876 of the Regio Ufficio Centrale di Meteorologia (UCM) – now CRA-CMA[1] – whose head office was located inside the ancient "Collegio Romano" (founded in the second half of the 16th century by the Jesuit Fathers), in the heart of Rome, where the astronomic and meteorological Observatory of the Pontifical State was run.

UCM founded a national monitoring network which rapidly grew from 50 to 400 stations in 1880. At the same time, (i) new rules to standardize measures and observations were defined and (ii) the systematic collection of meteorological data was activated. This latter is still in progress enriching the CMA Paper Archive.

Between the late nineteenth and the early twentieth century, thanks to specific international agreements, UCM started to gather meteorological data coming from many European and international weather stations, that, now, are still stored in suitable yearbooks.

---

[1] Office names and acronyms since 1876:

1876–1887 *Regio Ufficio Centrale di Meteorologia – UCM* (Royal Central Office for Meteorology)

1887–1923 *Regio Ufficio Centrale di Meteorologia e Geodinamica – UCMG* (Royal Central Office for Meteorology and Geodynamics)

1923–1939 *Regio Ufficio Centrale di Meteorologia e Geofisica – UCMG* (Royal Central Office for Meteorology and Geophysics)

1939–1941 *Regio Ufficio Centrale di Meteorologia e Climatologia – UCMC* (Royal Central Office for Meteorology and Climatology)

1941–1956 *Ufficio Centrale di Meteorologia e Ecologia Agraria – UCMEA* (Central Office for Meteorology and Agrometeorology)

1957–2007 *Ufficio Centrale di Ecologia Agraria – UCEA* (Central Office for Meteorology and Agrometeorology)

2007–today *Unità di Ricerca per la Climatologia e la Meteorologia applicate all'Agricoltura – CRA-CMA* (Research Unit for Climatology and Meteorology applied to Agricolture of the Agricultural Research Council).

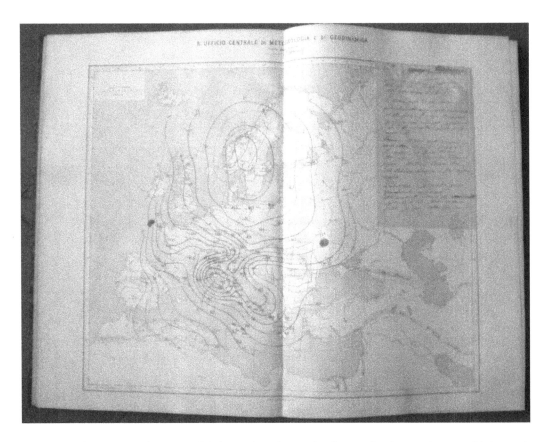

Figure 1. Sample of an early daily isobaric chart. A brief note about the weather forecast for the next day was added to each map.

The early UCM goals were: (1) to carry out studies on Italy weather and climate using the above-mentioned meteorological information and (2) to produce weather forecasts called "Presagi" (forecast of future events). These activities were carried out, gathering data from meteorological stations, updating weather charts and producing daily forecasts, broadcasted by telegraph throughout Italy and Europe.

These original maps and messages are still kept at the CMA Library (Fig. 1).

2 The CMA Paper Archive

The CMA Paper Archive (Fig. 2) collects about 3300 folders, each containing a yearly collection of ten-days observation forms, filled by volunteer observers, producing daily or sub-daily hand written records of air temperature, relative humidity, atmospheric pressure, wind speed and direction, precipitation, sunshine duration, solar radiation, evaporation and visual observations (cloud cover and type, snow depth and so on).

Some of the early observatories belonging to the UCM network are nowadays still working. The Archive holds more than 850 series. Among them, 260 are longer than thirty years, 20 are longer than one hundred years and 40 are still in progress at present.

For each standard meteorological variable the Archive has more than six millions elementary measurements coming from Italy or other countries like the former Italian territories of Balkans, the former Yugoslavia, Greece and Albania. Furthermore, meteorological data from the Italian Colonies of Africa (Ethiopia, Libya and Somalia) are also held in the CMA Library.

One section of the observation forms, named *Note speciali sui fenomeni meteorologici* (Special notes on meteorological phenomena), owing to remarks concerning special weather events, is written by the observer in his own words. Sometimes these remarks show colorful or poetic comments about atmospheric phenomena. Moreover, specific comments about local historical events can be read in these scientific forms, that, in this way, become noteworthy historical and cultural documents (Fig. 3).

The observational forms also report information about replacement of the observer, shift in reporting time, changes in site characters, failure and replacement of instruments, and so on. Those metadata allow us to trace the history of each weather station and provide a key for critical analysis of data, improving the validation and standardization processes.

As far as written above, the Paper Archive represents a valuable source of meteorological data of both past and present. CRA-CMA is spending great efforts to implement

Figure 2. The seat of the Meteorological Archive at CRA-CMA.

the digitization of paper data, in order to constantly update its electronic databank named National Agrometeorological Data Base (BDAN)[2]; digitization is a priority program for CMA and it will continue in the future. At present, (we assume that) less than 30 % of the whole original Paper Archive has been processed.

However, digitization only covers the main meteorological variables, as temperature, precipitation, pressure, sunshine, wind and leaves behind all auxiliary information like the witness remarks mentioned above, the visual observations, the type and duration of weather events and others metadata. Since scanning and image capture of paper forms are the best approach to preserve their full content, a program for the creation of an "Image-file Archive" is almost ready to start.

The CMA Archive keeps another relevant collection: the *Thunderstorms postcards* (Fig. 4) compiled and sent by several observers to the Office between 1880 and 1970. It represents a source of interesting information about these severe weather phenomena. Postcards are not fully cataloged, but a corpus of about ten thousand thunderstorms postcards is estimated.

Furthermore, between 1887 and 1939, UCM was charged to manage the National Seismic Service. In this role, it received information about all earthquakes occurred in Italy by special postcards (Fig. 5) sent from numerous geodynamic observatories and seismic stations located throughout the Country. The collection consists of about 5000 seismic postcards, cataloged by the National Institute of Geophysics and Volcanology (INGV) of the Italian National Research Council (CNR).

3 The CMA Library

The CMA Library (Fig. 6) digs its roots in the early years of the Collegio Romano Astronomical and Meteorological Observatory. In fact, the Library comes from the Earth Sciences collection of the early Observatory (Fig. 7) integrated with later meteorological and geodynamic books. In 1879 this bibliographic collection became the core of the UCM Library.

Thereafter, the history of the Library can be identified with the history of the Office. During the period in which UCM held the responsibility for the "National Geodynamic Service" (1887–1939), a lot of books and periodicals on seismology, volcanology and geology were acquired by the Library, either from Italy and abroad.

In 1939, when the geodynamic competences of UCMG were transferred to the ING (now INGV), some of the geophysical and geological collections were transferred to this institution. Afterwards UCM focused its activity on the fields of ecology and meteorology applied to agriculture. Hence, over the years, the CMA Library has been enriched with many valuable and rare collections of books and periodicals on meteorology, climatology and phenology, becoming the leading Italian Library in these fields.

The Library collects now more than 15 000 historical texts related to meteorology; most published in the nineteenth century and in the early twentieth century. Moreover, some historical books on geophysics date back to the second half of the sixteenth century.

The collection contains a large variety of books on synoptic and dynamic meteorology, such as historic texts by Louis Cotte, Giuseppe Toaldo, Matthew Fontain Maury, Elias Loomis, Angelo Secchi, Robert Fitz Roy, Urbain Le Verrier, Christophorus Buys Ballot, Adolphe Quetelet, Edme Hippolyte Marié-Davy, William Ferrel, Hermann von Helmholtz, Luigi De Marchi, Adolf Sprung, Wilhelm van Bebber, Ralph Abercromby, Léon Teisserenc de Bort, Alfred Angot, Francesco Vercelli, Vilhelm and Jacob Bjerknes, Lewis Fry Richardson, Jules Charney, Ernesto Gherzi.

Several of these books, relevant for climatology and climate change, constitute the written heritage of notable scientists, such as Giovanni Targioni Tozzetti (1712–1783), Vincenzio Chiminello (1741–1815), Jean B. J. Fourier, John Tyndall, Svante Arrhenius, Louis Agassiz, Luigi De Marchi, Filippo Eredia, Gustav Hellmann, Ellsworth Huntington, Guy S. Callendar, Milutin Milankovic, Gordon Manley, Hubert H. Lamb, Edward N. Lorenz, Ezio Rosini, Sabino Palmieri and Vittorio Cantù.

Among Italian periodical publications available in the Library, the most important Italian ones are listed below. Many of them were issued by the same Regio Ufficio Centrale di Meteorologia e Geodinamica (UCMG).

---

[2]Several datasets of the National Agrometeorological Data Base (BDAN) are available at the web site: http://www.cra-cma.it.

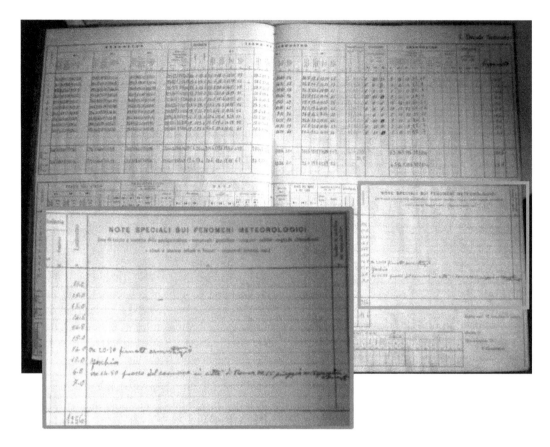

Figure 3. Collegio Romano Observatory: year 1943, September, the first ten days observation form. Weather notes and news about the Second World War are reported in "Special notes on the weather": 7 September, 08:30 p.m. – "The armistice" (Italy – Allied Countries) was signed; 8 September – Haze; 9 September, 02:50 p.m. "Cannon shots in Rome"; 03:00 p.m. "Rain"; 06:00 p.m. "Shooting in the Italian Senate".

– *Bullettino Meteorologico dell'Osservatorio del Collegio Romano* (Weather Bulletin of Collegio Romano's Observatory) (1862–1878) – issued to give visibility and access to meteorological data and Italian studies on meteorology and terrestrial physics. This magazine is the expression of the excellent work made by the Pontifical Meteorological Network founded by Jesuit Father Angelo Secchi in 1855, at the Collegio Romano Observatory, the first modern storms forecasting service in the world indeed.

– *Bullettino Meteorologico dell'Osservatorio del Real Collegio Carlo Alberto in Moncalieri* (Meteorological Bulletin of Moncalieri's Observatory) (1865–1923) – the first scientific publication of its kind in Northern Italy, similar to the Collegio Romano Bulletin. This periodical was the journal of a private meteorological network established by Father Francesco Denza, covering all the Italian regions (Corrispondenza Meteorologica Italiana delle Alpi e degli Appennini). This network operated in parallel with the UCM one.

– *Meteorologia Italiana* (Italian Meteorology) (1865–1878) – scientific magazine published by the Ministry of Agriculture, Industry and Commerce's Climatological Service. This first governmental periodical on meteorology and climatology mainly contains meteorological data. From 1867 a Supplement on the Italian meteorology (*Supplemento alla Meteorologia Italiana*) was added to the journal, issuing relevant meteorological analyses and original notes and memories on meteorological topics.

– *Annali del R. Ufficio Centrale di Meteorologia e Geodinamica* (Annals of the Royal Central Office for Meteorology and Geodynamics) (1879–1935) – issued by UCM, this collection of volumes is the prosecution of *Meteorologia Italiana*. Each yearly volume is subdivided in three parts. The first part contains memories and papers on meteorology and geodynamics, produced by the scientific staff of the UCM and other Italian meteorologists and geophysicists. Meteorological works are referred to several different topics:

Figure 4. Sample of a Thunderstorms postcard. Front: Thunderstorms Service – Meteorological Observatory Rome. Back: Province Avellino, Station of Avellino, Thunderstorm of Day 7, Month August, Year 1882. Start at hours 1 minutes 50 p.m. End at hours 04:35 p.m., id = Thunderstorm direction SW To NE. Wind direction SW. Rain Torrential. Hail = Remarks: thunderstorm starts with weak and protracted thunders far at W. Rain at the first time regular, about at 02:15 p.m. torrential. Wind of changeable intensity; in the maximum phase very strong. At 04:35 p.m.: thunderstorm stops and sky gradually becomes clear. Signature not understandable.

- researches on thunderstorms in Italy, authored by Schiaparelli, Frisiani and Pini,

- pioneering studies on theoretical and dynamic meteorology, mostly by Luigi De Marchi,

- contributions to the knowledge of Italian Climate, by Filippo Eredia and Elia Millosevich,

- pioneering memories on the emerging field of Aerology (Aerological Observatory of Pavia),

- researches about terrestrial magnetism in different Italian regions by Ciro Chistoni and Luigi Palazzo.

Figure 5. Sample of a Seismic postcard. Front: to Central Office for Meteorology and Geodynamics In Collegio Romano Via del Caravita, 7 Roma. Back: earthquake occurred at Badia Calavena District = Province Verona, Day 21, Month August, Year 1891, hours 11:00 p.m. min 30 sec. ... Very powerful sussultatory quake, but not like those of 09:00 p.m. and min. ?. After this [...] soil quaked without interruption and we counted 23 small quakes up to 03:00 'o clock in the afternoon of the following day 22. The quake at 09:00 p.m. of 21st brought panic and consternation among the people who abandoned houses. Many damages to buildings, especially in quarters Cessari and Minazzi of the same town. Observer name: A. Govran.

The remaining parts of the *Annali* are devoted to meteorological data recorded by the UCM network.

- *Bollettino Meteorico Giornaliero* (Daily Weather Bulletin) (1879–1927) – Published by UCM, it was issued since 30 November 1879 in order to gather daily weather data from main Italian and foreign observatories. On 1 August 1880 the number of contributing stations was substantially increased to provide a summary

Figure 6. The Library – "the Tacchini Gallery".

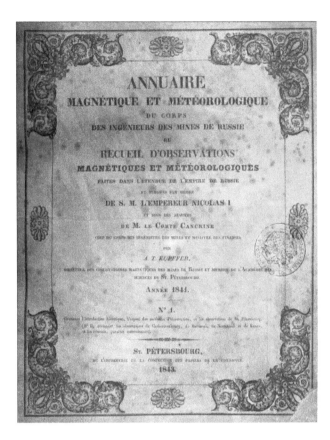

Figure 7. Magnetic and Meteorological Yearbook, St. Petersburg, 1841.

of weather conditions in Italy and Europe, as well as a specific forecast dispatch ("Presagio"). Further additions and modifications gave to the journal the final form of a four sided lithographed sheet with daily isobaric and isothermal charts of the Mediterranean Region and one European isobaric chart. A daily broadcast of the "Presagio" was carried out by telegraph to main Italian observatories and harbourmasters (from 1 April 1927, this kind of broadcast was prosecuted by the Weather Office of the Air Force Ministry).

Rivista Meteorico-Agraria Decadica (Ten-days Agrometeorological Magazine) (1879–1916) – published by UCM, it showed, for a ten days period, the European barometric situation, the weather conditions in Italy, such as extreme temperatures, cloudiness, precipitation, etc., offering an interesting overview of crop growth and development for the different agricultural areas of our Country. Each magazine was improved with a map showing average isotherms and the rainfall distribution in Italy. Short agrometeorological notes were also published sometimes as an appendix.

Many Italian and foreign periodicals on Aerology, Oceanography, Geography, Geology, Seismology, Volcanology, as well as many early weather charts from the second half of the 19th centuries of both historical and scientific interest are also stored in the CMA library. Moreover, the Library contains numerous conference papers, working reports and monographs produced by UCM researchers and employees since 1879. It should also be noted that the Library holds some original manuscripts, unpublished correspondences (19th and 20th centuries) and some interesting miscellaneous collections, e.g. more than fifty volumes containing memoirs and papers on Meteorology and Geophysics, written by the greatest Italian scientists of the 19th and 20th centuries, such as Domenico Ragona, Pietro Tacchini, Luigi Palazzo, Giuseppe Mercalli,

Mario Baratta, Alfonso Cavasino, Ludovico Marini, Emilio Oddone, etc.

Over the years, the Library consultation has been a valuable aid to many experts and researchers in geophysics and meteorology to produce original studies, leading to the redaction of new books, articles, essays and academic works.

It has been recently learned that among the scholars who attended the Library, there was the young Enrico Fermi (1901–1954). During the summer of 1918 Fermi attended his studies at the Library to prepare for the admission exams of Scuola Normale Superiore di Pisa. A key step in his preparation was the study of a French edition of the Treaty of Physics by Orest D. Chwolson, professor at the Imperial University of St. Petersburg. In Rome only two copies of this work were probably available: one in the Library of the Institute of Physics at via Panisperna, not consultable by simple students like Fermi, while the other copy was and is still held in the CRA-CMA Library. Filippo Eredia (Fermi's teacher of physics at the school "Liceo Umberto" in Rome and at the same time responsible for the "Servizio Presagi" at UCM) made the book available to Fermi, which spent most of that summer immersed in reading this Treaty in the CMA Library.

At present, the CMA Library is known as the "Central Library of the Italian Meteorology", because it is the largest Italian collection in atmospheric sciences, the main Italian memory of the historical tradition for Meteorology and one of the main bibliographic depository for Geophysics.

The Library is about to join the National Library Service (SBN), in order to get more visibility and make its heritage more accessible to researchers.

4 Concluding remarks

The Library and Paper Archive of CRA-CMA are a rich repository of historical meteorological data. Climatic information is insufficiently known and by consequence largely under-exploited by the scientific community. On this basis, CMA's main goals are (i) to give and ease access to the large variety of the historical weather data of the Paper Archive and Library and (ii) also to raise credit abroad and (iii) become a fruitful authorship for the international research on historical climatology.

Several projects have been planned in order to achieve a significant exploitation of these data. More specifically, the ongoing digitizing program represents a first step to draw up at European scale the following climatic studies, also useful to evaluate the impact of climate change on growing crops:

- analysis of climate dynamics in a comparative historical perspective, with special regard to the Euro-Mediterranean area;

- recognition of daily atmospheric circulation patterns at sub-synoptic scale and mesoscale;

- spatialisation of secular temperature and pressure fields at mesoscale;

- generation of daily datasets, useful, for example, as input for phenological or crop production models;

- generation of datasets of phenological observations useful for phenological models validation.

At present CMA is acknowledged in Italy as an important institution providing reference and consulting services on meteorological and climatic data to Italian universities, agencies and other national institutions.

This presentation would enhance the information exchange and achieve new international relationships and co-operations.

Acknowledgements. The Authors are very grateful to Luigi Mariani and Gabriele Cola for the linguistic revision of this paper.

Edited by: I. Auer
Reviewed by: M. Maugeri and another anonymous referee

13

Has the frequency or intensity of hot weather events changed in Poland since 1950?

J. Wibig

Department of Meteorology and Climatology, University of Lodz, Poland

Correspondence to: J. Wibig (zameteo@uni.lodz.pl)

Abstract. Various indices of hot weather frequency and intensity were analysed in the area of Poland in the period between 1951 and 2006. An increase of majority of them was shown in the whole year and all summer months but September, when significant decrease in all indices was apparent. The correlation of selected hot weather indices and precipitation totals in a month of hot weather event and the preceding months were also calculated to check if prolonged dry weather can constitute a forcing factor for hot event creation. Because significant correlations appear mainly in the cases when precipitation is for the same month as the hot weather index, it seems that in Poland the presence of high pressure systems is a more important factor of hot event creation than dry weather.

1 Introduction

There is increasing evidence that the frequency and intensity of extreme weather events have changed considerably since the middle of the twentieth century (Trenberth et al., 2007). Moberg and Jones (2005) have shown that all tails of minimum and maximum temperature distribution (defined on the percentile basis) have increased since the beginning of the twentieth century. The increase in upper tail of daily maximum and minimum temperatures in summer relates to the rise in frequency and intensity of hot events. Extreme hot events exert strong social, economic and environmental impacts. They can cause the death of people, forest fires and energy losses. Besides climate warming, hot events are believed to become more frequent, intense and prolonged (Beniston, 2004; Schär et al., 2004). So it is important to find a mechanism triggering their occurrence and/or to link it to atmospheric circulation. Della-Marta et al. (2007) have shown the relationship of heat waves over western Europe to large scale forcings and the anomalous high pressure systems over Scandinavia and central western Europe. Garcia-Herrera et al. (2010) connected occurrence of prolonged hot weather with quasi-stationary anticyclonic circulation anomalies. It corroborates the results of Wibig et al. (2009) that heat waves in Poland are accompanied by

higher than normal pressure over central and eastern Europe, with the strongest positive pressure anomalies located just in the east of Poland. Some authors have found that heat wave occurrence can be influenced by precipitation deficit and pre-existing dry soil conditions (Della-Marta et al., 2007; Fischer et al., 2007). The aim of this paper is to check if prolonged dry weather can constitute a forcing factor for hot event creation in Poland.

2 Data and methods

Daily maximum temperatures from 21 stations from Poland (Fig. 1) from the period 1951–2006 were used for describing the hot event frequency and intensity. Monthly precipitation totals from the same stations and period were used for analysis of the influence of drought events on hot event development. The homogeneity of records of mean monthly maximum temperature and monthly precipitation totals were tested with Standard Normal Homogeneity Test by Alexandersson (1986).

The hot event frequency was assessed by means of monthly and annual number of days with $T_{max} > 25\,°C$ and $T_{max} > 30\,°C$. The hot event intensity was characterized by monthly and annual sum of T_{max} exceeding $25\,°C$ and $30\,°C$ as well as by the annual number of days with T_{max} above

Figure 1. Location of IMWM stations with data on monthly precipitation totals and daily maximum temperature (1951–2006).

95 and 99 percentiles of T_{max} in the reference period 1961–1990, the longest spell of days with $T_{max} > 30\,°C$ in the year and the highest sum of $T_{max} > 30\,°C$ on all days within one such spell.

The trends of these indices were calculated using linear regression with parameters obtained with the least square method and statistical significance tested by Student t-test. Impact of droughts on hot weather frequency was tested correlating precipitation totals in one, two and three months preceding the specific month with monthly number of days with $T_{max} > 25\,°C$ and $T_{max} > 30\,°C$ and with monthly sum of degree-days exceeding $30\,°C$ on this month.

3 Results

3.1 Hot weather occurrence and its long-term variability

In lowland Poland, days with $T_{max} > 25\,°C$ occur mainly from May to September, but in extreme cases they can appear also in April and October. An example from Legnica is presented in Fig. 2, but similar pictures can be obtained for all analysed station, but a few located at the sea side or in the mountains. There are more than ten such days in July and August, 7–8 in June and about 3 in May and September. The number of degree-days exceeding $25\,°C$ vary from 0 to about 150 during extremely warm months. Days with $T_{max} > 30\,°C$ occur mainly in July and August, but they can happen also in May, June and September. There are about five such days in July and August and only about one in July. The number of degree-days exceeding $30\,°C$ vary from 0 to about 50 during extremely hot months.

Since 1951, the number of days with $T_{max} > 25\,°C$ has increased in Poland in the period 1951–2006. The increase was the strongest in August, when it was statistically signif-

icant at more than 70 % of analysed stations (Fig. 3). An increase was also observed in July, May and in the year as a whole. A decreasing trend at 16 stations out of 21, and significant at 7 stations, was observed in September. In the case of days with $T_{max} > 30\,°C$ significant increasing trends were observed at more than 60 % of stations in July and about 30 % of stations in June. A significant decreasing trend was detected at about 45 % of stations in September. The number of degree-days exceeding $25\,°C$ has increased significantly at more than 60 % of station in July and more than 50 % of stations in August. At the same time, the number of such days has decreased significantly at more than 50 % of stations in September. The number of degree-days exceeding $30\,°C$ has increased significantly at almost 50 % of stations in July, at about 35 % of stations in June and about 20 % of stations in May and August. At the same, time the number of degree-days has decreased significantly at 35 % of stations in September.

An example of long-term course of the annual number of days with temperature exceeding 95 and 99 percentile for the reference period 1961–1990 is shown in Fig. 4. The increasing trend of both records is well seen. Statistically significant trends in both records were observed at more than 80 % of analysed stations.

Figure 5 presents an example of long-term course of the longest annual spell of consecutive days with $T_{max} \geq 30\,°C$ and the most intensive such spell in degree-days. Graphs for other stations are similar. The maximum values in extremely hot years 1994 and 2006 are well seen. Positive linear trends characterise both records at the majority of stations, but because the distribution of data is far from normal, the significance of these results cannot be assessed using a simple t-Student test.

3.2 Relation of hot event occurrence to pre-existing precipitation deficit

To analyse the influence of prolonged drought on hot event occurrence, the precipitation totals in one, two or three months with monthly number of days with $T_{max} > 25\,°C$ were calculated. In one version, the last month of precipitation data agrees with the month of hot event index (i.e. precipitation totals from March, April and May with the number of days with $T_{max} > 25\,°C$ in May); in the second, the last month of precipitation data precedes the month of heat wave index (i.e. precipitation totals from March, April and May with the number of days with $T_{max} > 25\,°C$ in June). In the first case, significant correlations at more than 50 % of stations occurred in all cases (precipitation totals from one, two or three months) from June to September (Table 1). In the second case, only in September significant correlations appear at more than 30 % of stations. It means that the month with a hot event has to be simultaneously dry. Because the quasi-stationary high pressure system located over Poland or in its neighbourhood corresponds both to hot weather (Wibig

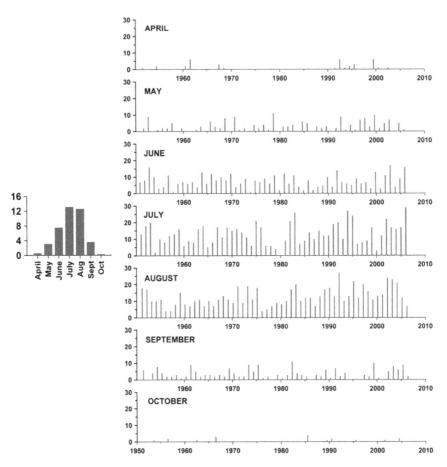

Figure 2. The intraannual distribution (left) and the long-term course of the number of days with $T_{max} > 25\,°C$ in the period 1951–2006 (right). An example from Legnica.

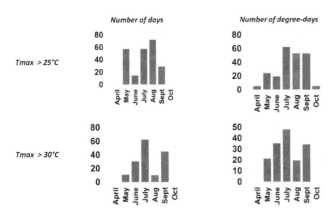

Figure 3. The percentage of stations with significant increasing (blue) and decreasing (red) trends.

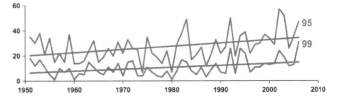

Figure 4. Number of days with maximum temperature exceeding 95 and 99 percentile according to the reference period 1961–1990 based on the example of Kalisz.

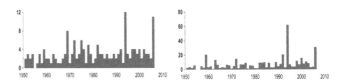

Figure 5. The longest annual spell of consecutive days with $T_{max} \geq 30\,°C$ (left) and the most intensive spell in degree-days (right), an example from Slubice.

et al., 2009) and lower than normal precipitation (Piotrowski, 2010) it appears that in Poland it is a much more important factor in creation of hot events than dry weather.

On persistent blocking high pressure system as a trigger of heat wave indicated also Beniston and Diaz (2004) for Switzerland, Xoplaki et al. (2003) for Greece and Carril et

al. (2008) for the north-western Europe and Eurasia (2008). However according to some investigations dry soils in spring

Table 1. The number of stations with correlation coefficients between the precipitation totals and monthly number of days with $T_{max} > 25\,°C$ significant at 5 % level.

	April	May	June	July	August	September	October
	last month of the period of precipitation total in line with that of the index						
1 month	1	1	14	21	15	17	1
2 months	1	2	14	20	17	19	0
3 months	1	1	11	18	15	17	0
	last month of the period of precipitation total precedes that of the index						
1 month	4	1	2	2	2	9	0
2 months	4	1	2	2	3	7	0
3 months	5	1	1	1	3	11	2

over over Mediterranean areas favor and maintain hot summers in this region (Vautard et al., 2007).

4 Summary

All analysed indices of hot event frequency and intensity have shown an increasing trend in July, August and the whole year, in some cases also in June. At the same time, all hot weather indices in September have decreased. An increase in hot event frequency and intensity is still insignificant in the eastern part of Poland.

There is a strong relation of hot weather occurrence with precipitation (dryness). This relation is significant when the periods of precipitation and hot event occurrence overlap. It means that a month with a higher than average number of hot days simultaneously has precipitation below average. A question arises: is it a direct effect of precipitation lack or of the fact that in summer high temperature and low precipitation occur when the quasi-stationary high pressure system is located over Poland or in its neighbourhood?

Acknowledgements. The author thanks the IMWM for kindly providing the data and has been funded by the project Nr. 0533/B/P01/2009/36 of MS&HE.

Edited by: I. Auer
Reviewed by: two anonymous referees

References

Alexandersson, H.: A homogeneity test applied to precipitation data, Int. J. Climatol., 6, 661–675, 1986.

Beniston, M.: The 2003 heat wave in Europe: a shape of things to come? An Analysis based on Swiss climatological data and model simulations, Geophys. Res. Lett., 31, 1–4, 2004.

Beniston, M. and Diaz, H. F.: The 2003 heat wave as an example of summers in a greenhouse climate? Observations and climate model simulations for Basel, Switzerland, Global Planet. Change, 44, 73–81, 2004.

Carril, A. F., Gualdi, S., Cherchi, A., and Navarra, A.: Heatwaves in Europe: areas of homogeneous variability and links with the regional to large-scale atmospheric and SSTs anomalies, Clim. Dynam., 30, 77–98, 2008.

Della-Marta, P. M., Luterbacher J., von Weissenfluh, H., Xoplaki E., Brunet, M., and Wanner, H.: Summer heat waves over western Europe 1880–2003, their relationship to large-scale forcings and predictability, Clim. Dynam., 29, 251–275, 2007.

Fischer, E. M., Seneviratne, S. I., Vidale, P. L., Luthi, D., and Schar, C.: Soil moisture – Atmosphere interactions during the 2003 European summer heat wave, J. Climate, 20, 5081–5099, 2007.

Garcia-Herrera, R., Diaz, J., Trigo, R. M., Luterbacher, J., and Fischer, E. M.: A Review of the European Summer Heat Wave of 2003, Crit. Rev. Env. Sci. Tec., 40, 267–306, 2010.

Moberg, A. and Jones, P. D.: Trends in indices for extremes of daily temperature and precipitation in central and western Europe 1901–1999, Int. J. Climatol., 25, 1173–1188, 2005.

Piotrowski, P.: The circulation conditioning of the spring droughts in Poland, in: Klimat Polski na tle klimatu Europy. Warunki termiczne i opadowe, edited by: Bednorz, E., Bogucki wyd. Nauk. Poznañ, 147–158, 2010 (in Polish).

Schär, C., Vidale, P., Luthi, D., Frei, C., Haberli, C., Liniger, M., and Appenzeller, C.: The role of increasing temperature variability in European summer heatwaves, Nature, 427, 332–336, 2004.

Trenberth, K. E., Jones, P. D., Ambenje, P., Bojariu, R., Easterling, D., Klein Tank, A., Parker, D., Rahimzadeh, F., Renwick, J. A., Rusticucci, M., Soden, B., and Zhai, P.: Observations: Surface and Atmospheric Climate Change, in: Climate Change 2007: The Physical Science Basis. Contribution of Working Group I to the Fourth Assessment Report of the Intergovernmental Panel on Climate Change, edited by: Solomon, S., Qin, D., Manning, M., Chen, Z., Marquis, M., Averyt, K. B., Tignor, M., and Miller, H. L., Cambridge University Press, Cambridge, United Kingdom and New York, NY, USA, 2007.

Vautard, R., Yiou, P., D'Andrea, F., de Noblet, N., Viovy, N., Cassou, C., Polcher, J., Ciais, P., Kageyama, M., and Fan, Y.: Summertime European heat and drought waves induced by wintertime Mediterranean rainfall deficit, Geophys. Res. Lett., 34,

L07711, doi:10.1029/2006GL028001, 2007,

Wibig, J., Podstawczyñska, A., Rzepa, M., and Piotrowski, P.: Heatwaves in Poland – frequency, trends and relationship with atmospheric circulation, Geographia Polonica, 82, 33–46, 2009.

Xoplaki, E., González-Rouco, J. F., Gyalistras, D., Luterbacher, J., Rickli, R., and Wanner, H.: Interannual summer air temperature variability over Greece and its connection to the large-scale atmospheric circulation and Mediterranean SSTs 1950–1999, Clim. Dynam., 20, 537–554, 2003.

14

Socio-economic benefits of weather and climate services in Europe

A. Perrels[1], Th. Frei[*,], F. Espejo[2], L. Jamin[3], and A. Thomalla[3]**

[1]Finnish Meteorological Institute FMI, Erik Palménin Aukio 1, P.O. Box 503, 00101 Helsinki, Finland
[2]Spanish State Meteorological Agency AEMET, C/Leonardo Prieto Castro 8, 28040 Madrid, Spain
[3]Deutscher Wetterdienst DWD, Frankfurter Str. 135, 63067 Offenbach, Germany
[*]formerly at: MeteoSwiss, Krähbühlstrasse 58, P.O. Box 514, 8044 Zürich, Switzerland
[**]presently: private consultant (thomas.frei1@bluewin.ch)

Correspondence to: A. Perrels (adriaan.perrels@fmi.fi)

Abstract. There is a rising interest around the world for a better understanding of the economic and social value added of weather services. National hydro-meteorological services and international cooperative bodies in meteorology have ever more to justify their use of public budgets. Furthermore, the development of hydrological and meteorological services is to a large extent steered by expectations regarding the eventual benefits of the envisaged new developments. This article provides a compact overview of the impediments for uptake of socio-economic benefit (SEB) studies, methods and results of SEB studies to date. It also discusses some pitfalls and crucial steps to enhance a broader uptake of SEB studies.

1 Introduction

The national meteorological and hydrological services (NMHSs) in Europe have been steadily improving their forecast skills and expanding their environmental and climate services (Nurmi and Brockmann, 2007), but face nonetheless major challenges. Public budgeting is tight and only seems to get tighter. A growing range of media through which weather information is provided blurs the position of NMHSs in public and commercial weather service provision. Varying interpretations of free access to public information and competition promotion in weather service markets create uncertainty for the future scope of NHMSs, while technical and scientific progress create new resource demands as well as new service opportunities (World Bank, 2008).

To date all reviews of Socio-economic benefits (SEB) of hydro-meteorological services show very good benefit-cost ratios, i.e. in any country the benefits of these services for society are many times their costs (see Sect. 4). Yet, the sense of doing SEB studies from an NMHS perspective has more to do with systematic learning than with marketing splendid *B/C* ratios. The Task Team Socio-Economic Benefits (TT

SEB) of the WMO Working Group on Public Service Delivery will soon publish a report on social-economic benefit assessment of weather services in which is explained what can be the purposes and approaches and how a SEB study can be organised. The main message of that report is to "*grasp the benefit of showing the benefits*". In this article, based on that report, we will briefly discuss impediments, methods and pitfalls, and results.

2 Developments in Europe

In the past ten years the WMO has been running several programmes with respect to SEB of hydro-meteorological services (WMO website, 2013). The 2007 WMO Madrid International Conference has been an important catalyst for interest in SEB studies. In September 2011 a survey was carried out regarding the interest for SEB assessment among WMO Regional Association Europe (RA VI) members. A summary of the results was presented at the WMO RA VI Conference on Social and Economic Benefits in 2011 in Lucerne, Switzerland (WMO, 2012).

Interest was wide spread, but so far only a minority has been carrying out such assessments. Lack of in-house knowledge on SEB studies, uncertainty regarding how to commission a meaningful SEB study, and doubts about adequate resourcing of SEB studies appeared to be important impediments.

The survey was carried through internet and addressed to the 50 WMO RA VI region member NMHSs. Answers from 25 NMHSs were received. The results of the survey denoted that:

- 21 (i.e. 84 %) of the responding NMHSs carry out (more or less regularly) user group surveys, such as fora or user consultations, which can be seen as a good start to map user needs and benefits.

  - 12 of the responding NMHSs (i.e. 57 % of those running surveys) try to estimate the user benefits regarding their meteorological services

- only 7 NMHSs (i.e. 28 % of the respondents) indicated that they have already experiences with SEB studies; among those 7 NMHSs 4 used cost-benefit based methods.

On the other hand 17 (43 %) of the responding NMHSs indicated that they plan to carry out SEB studies for several reasons, such as (it was allowed to tick more than one option):

- justify funding from the government (17);

- promote its services and attract new users (14);

- use results for decision support to prioritise their investment (11).

Out of the 25 responding organisations 17 NMHSs indicated to be interested in receiving additional guidance and advise on appropriate methodology. In addition, the respondents expressed their interest in sharing experience among NMHSs as regards SEB studies as well as related funding opportunities, e.g. through WMO supported projects.

3 Methods

There is no unique method for assessing the social and economic benefits generated by a NMHS; it depends on the field of the activity within the NMHS (weather forecasting, climate services, warnings and emergency services, other environmental products, etc.) and on the sectors to which these services are oriented. Different market conditions affect the range of applicable methods, whereas also the precise questions to be answered has significant impact on what is a useful method. Any method chosen should somehow be capable of (1) identifying how the use of weather services makes a difference in the costs or revenues of a sector or company and (2) of estimating the size of the benefit attributable to

the weather service by means of a formalized description of the differential effect on costs and/or revenues. Even though there is quite a choice of conditionally relevant methods, some general considerations can be made, such as outlining the steps along which the value added is created by the supply and use of weather services.

The creation of value added (VA) in the supply chain for weather services occurs in three phases (Fig. 1), being: (VA1) when combining data, models and expertise to generate weather forecasts and adjacent services, (VA2) when editing and distributing weather information through media channels and enabling the combination of information, and (VA3) when end-users interpret weather information and use it in decision making (to avoid damages and exploit weather related opportunities). The first phase of value added creation (VA1) represents the activities in the NHMSs and commercial weather service providers. VA2 represents the different media channels (TV, radio, newspapers, websites, mobile), which convey weather information alongside other information.

The third phase (VA3), in which the end-users use weather information to avoid damage or reap opportunities, represents by far the largest amount of benefits, i.e. $VA3 \gg VA1 + VA2$. The extent to which the potential value of VA3 is realized depends on the personal or collective capacity of the end user to interpret and use that information adequately.

The above explanation was intently focused on weather services. For traditional climate services (such as for hydropower and agriculture) the model as represented in Fig. 1 largely holds, even though such services are usually tailored for specific customers and not general public service. The newly emerging climate services aimed at adaptation support are as yet much less standardized and have also a larger variety of suppliers, whereas also the nature of the service can be quite different. As a consequence Figs. 1 and 2 are not necessarily straightaway applicable to adaptation oriented climate services, while the applicable evaluation methods are quite different as well, among others because the evidence of the benefits is either a projection or based on more rapidly realized side benefits, such as avoided road maintenance cost owing to new observation based asset management systems.

As stated above valuating the SEB of weather services requires the assessment of the *differential effect* of (improved) weather information on the value added of a sector, i.e. the difference in generated benefits owing to a difference in information level. It is important to understand that generating benefits through meteorological services depends on how this information is delivered, understood and how user decisions are taken. The information delivery process can also be understood as a process of *information decay* during subsequent steps of information distribution and uptake. Information decay refers to the fact that in subsequent stages of the weather service chain, some of the original value potential gets unattainable because timely access, use, comprehension,

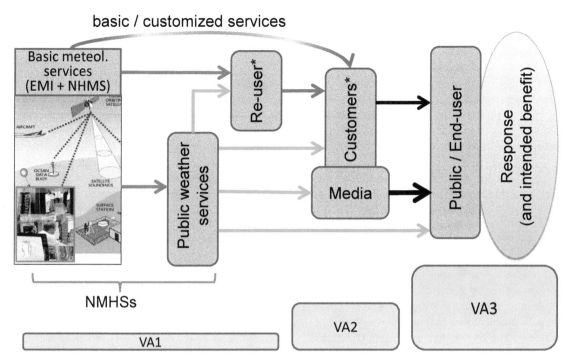

*) notably, but not exclusively, commercial weather service providers
**) includes both commercial *and* public sector customers

Figure 1. Principal stages in the provision of meteorological services and the related generation of value added by stage; VA = value added (by service stage 1, 2, 3).

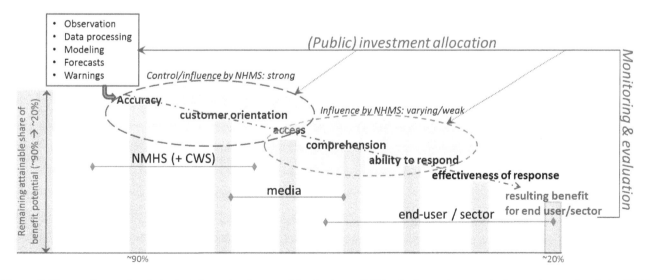

Figure 2. The weather service chain, its information decay and feedback possibilities for improving steps in the weather service chain; CWS denotes "commercial weather services".

etc. are not optimal. So, even the original forecast accuracy is 90 %, eventually only 20 % of the original value potential may get realized as the other steps (timely access, etc.) are not perfect (i.e. fractions varying between 80 % and 40 %, culminating in a compound fraction of 20 % or lower).

In many cases a hypothetical benefit potential (i.e. avoided cost based on ideal conditions) can be estimated thanks to available statistics on accidents, damage claims, etc. Subsequently, the degree of information decay in the weather service chain (Fig. 2) can be analysed so as to approximate the current level of realized benefits in the considered sector. Information on the sources of information decay can be used to improve steps in the weather service chain. The approach, denoted as weather service chain analysis (WSCA; Nurmi et al., 2012), can be quite generally applied in combination with various methods, including so-called Cost-Loss analysis which has been widely applied in appraisal studies in the weather service sector (Katz and Murphy, 1997). Essentially WSCA is an analysis of a product sum (eventual share of the benefit potential realized) and its constituent arguments (fractions). The consecutive arguments (fractions) can be assessed by means of underlying estimated functions, with variables representing abilities, propensities or proneness with respect to the considered fraction (Nurmi et al., 2012; Perrels et al., 2012).

4 Results

Studies in various countries, most of them industrialized, have shown that the cost/benefit ratio of a national meteorological service is around 1 : 3 to 1 : 10 as described in Table 1. However, there are only a few more or less comprehensive studies available and in many studies cost of media and of the end-users are not considered. Some studies have been estimating avoided cost (or deviation from maximum revenue), which relates to *realized* value added in an economy. Such an approach is useful if one wishes to answer how beneficial the weather services (and its public funding) actually have been for the national economy. In practice it means that the effect on value added of a bundle of closely related weather services used in a certain sector is analysed. The incremental contribution of one specific service product (via one particular media) in a considered bundle is often hard if not impossible to assess in this approach.

Other studies are based on willingness-to-pay (WTP) analysis. WTP studies give an indication of the potential market value of a particular weather service product, as they indicate the propensity to pay for the service in relation respondent's characteristics as well as the share of the potential user population which is prepared to pay a positive price. WTP studies can be applied to existing public services that are available free of charge (even though their use will require some kind of effort) as well as to new – not yet provided – services.

In the latter case the product characteristics and its potential usefulness should be unambiguous to the respondents.

5 Discussion

We are still far away from having a comprehensive understanding of social economic benefits of weather and climate services within Europe, but the results are so far encouraging. The available study results indicate – practically unanimously – favourable cost-benefit ratios. However, in many cases these ratios represent the overall bundle of public and private weather services taken together. For the NMHSs it would be helpful to clarify also what are the contributions of basic meteorological services, as well as what are the benefit-cost ratios of public weather services. In the WMO TT-SEB study (Perrels et al forthcoming) is in fact emphasized that the demonstration of good benefit-cost ratios should not be seen as the main output of SEB studies. Instead, SEB studies should be seen as a valuable source of information regarding (1) improvement of the effectiveness of current weather services in all phases of the weather service chain and (2) prioritizing focal areas of innovation of weather services. This does require however that SEB studies are repeated from time to time and are supported by regular monitoring for service use and user satisfaction. In turn this also implies an integration of the SEB studies, monitoring and exploitation into the management cycles of the NMHS.

The survey of 2011 and also the abundant attendance of the 2011 SEB conference showed that there is a wide spread interest for SEB studies. At the same time the lack of in-house knowledge seems to be a significant obstacle for getting a useful SEB study commissioned. Therefore the sharing of information and experiences across WMO members merits to be promoted, both with respect to carrying out such studies and regarding the uptake of the SEB study outcomes in the NMHS organisation. Twinning and pilot projects – in a WMO framework – for countries and NMHSs with a need for SEB studies but lack of know-how will also help to stimulate wider application of SEB studies. These twinning and pilot projects taken together should cover a wide variety of services and issues, such as (1) the sensitivity of various sectors as well as national economies as a whole, (2) the direct and downstream impacts of the various kinds of natural disasters, from short-lived storms to protracted droughts and (3) the likely social and economic impacts of long-term climate change on societies and economies.

Another important factor for greatly enhancing the societal value of, and benefits from, meteorological services is the establishment of much closer dialogue and sense of partnership between the provider and user communities at all levels.

There are still challenges with respect to applying the different methods, not the least due lacking data. In this respect a better monitoring of customers of weather services would help. This would be also a first advice to NMHSs that

Table 1. Overview of published cost-benefit analysis within different weather services, based on avoided costs unless otherwise stated.

Country	economic sector	cost-benefit ratio	Reference
Australia	general public (willingness to pay analysis) in Sydney	1 : 4	Anamann and Lellyett (1996)
Croatia	overview of all sectors	1 : 3 (at least)	Leviäkangas et al. (2008)
Denmark	overview of all sectors, 3 case studies	Many detailed examples with good net benefits; aggregate picture lacking	The Ministry of Transport and Energy (2006)
Finland	overview of all sectors	1 : 5	Leviäkangas and Hautala (2009)
	transport	1 : 10 and higher	Nurmi et al. (2012) internal FMI study (on-going for other sectors)
Nepal	mainly agriculture; transport and hydropower also considered	around 1 : 10	Perrels (2011)
Russia	overview of all sectors	1 : 3–1 : 4	Bedritsky and Khandozko (2001)
Switzerland	overview, all sectors	1 : 5	Frei (2010)
	transport sector	1 : 10	Frei et al. (2012)
United Kingdom	general public (willingness to pay analysis)	1 : 7	Met Office (2007)
	meteorological infrastructure – satellite	1 : 5–1 : 20	Joo et al. (2011)
USA	transport sector (winter road maintenance)	1 : 2–1 : 3	Ye et al. (2009)
	general public (willingness to pay analysis)	1 : 6	Lazo et al. (2009)

contemplate the commissioning of SEB studies. By means of user surveys a NHMS should first obtain an impression of how, for what precise purpose, and to what extent various user groups use the weather information. Such a survey assists to focus the SEB study and provides indispensable information to conduct it.

Some appraisal methods need further development to make them particularly suitable for economic assessment of weather and (traditional) climate services. This calls for intensifying co-operation with economic researchers. Furthermore, as regards the newly emerging adaptation oriented climate services standardisation of its economic evaluation practices is called for, but this requires also fundamental economic research work considering that a good part of the targeted benefits is supposed to realize in the (distant) future. This feature differs greatly from weather services or traditional (i.e. seasonal) climate services, for which evidence is produced within a year and partly within a week.

Acknowledgements. The comments of two anonymous reviewers and of the guest editor are kindly acknowledged. We also acknowledge the valuable feedback of Haleh Kootval and Dmitar Ivanov of the WMO and of Gerhard Adrian (director DWD) and Frits Brouwer (director KNMI) during the drafting of the SEB report, on which this article is based.

Edited by: H. Tuomenvirta
Reviewed by: two anonymous referees

References

Anaman, K. A. and Lellyett, S. C.: A Contingent Valuation Study of the Public Weather Service in the Sydney Metropolitan Area, Economic Papers: A Journal of Applied Economics and Policy, 15, 64–77, 1996.

Bedritsky, A. I. and Khandozko, L. A.: The economic benefits of hydrometeorological services, WMO-Bulletin, 50, 219–224, 2001.

Frei, Th.: Economic and social benefits of meteorology and climatology in Switzerland, Meteorol. Appl., 17, 39–44, 2010.

Frei, Th., von Grünigen, S., and Willemse, S.: Economic benefit of meteorology in the Swiss road transportation sector, Meteorol.

Appl., doi:10.1002/met.1329, in press, 2012.

Joo, S., Eyre, J. R., and Marriot, R. T.: The impact of Metop and other satellite data within the Met Office global NWP system, using and adjoint-based sensitivity method, Forecasting Research Technical Report 562, Met Office, UK, 2011.

Katz, R. W. and Murphy, A. H. (Eds.): Economic Value of Weather and Climate Forecasts, Cambridge University Press, Cambridge, 1997.

Lazo, J., Morss, R., and Demuth, J.: 300 Billion Served – Sources, Perceptions, Uses, and Values of Weather Forecasts, B. Am. Meteorol. Soc., 90, 785–798, 2009.

Leviäkangas, P. and Hautala, R.: Benefits and value of meteorological information services – the case of the Finnish Meteorological Institute, Meteorol. Appl., 16, 369–379, 2009.

Leviäkangas, P., Hautala, R., Räsänen, J., Öörni, R., Sonninen, S., Hekkanen, M., Ohlström, M., Venäläinen, A., and Saku, S.: Benefits of Meteorological Services in Croatia, Research Notes 2420, Finnish Meteorological Institute, Helsinki, 2008.

Met Office: The Public Weather Service's contribution to the UK economy, Met Office UK, www.metoffice.gov.uk/media/pdf/h/o/PWSCG_benefits_report.pdf, 2007.

Ministry of Transport and Energy (Denmark): Meteorology – a Revenue Generating Science; A mapping of meteorological services with an economic assessment of selected cases, Ministry of Transport and Energy, Copenhagen, http://www.wmo.int/pages/prog/amp/pwsp/documents/Meteorology_Revenue_Generating_Science.pdf, 2006.

Nurmi, P. and Brockmann, M.: A quarter century of operational end forecast verification, EMS Annual Meeting Abstracts, Vol. 4, FW4, P0102, EMS2007-A-00345 (San Lorenzo de el Escorial, Spain), http://meetings.copernicus.org/ems2007, 2007.

Nurmi, V., Perrels, A., Nurmi, P., Seitz, D., Michaelides, S., Athanasatos, S., and Papadakis, M.: Economic value of weather forecasts on transportation – Impacts of weather forecast quality developments to the economic effects of severe weather, EWENT report D5.2, http://ewent.vtt.fi/Deliverables/D5/D5_2_16_02_2012_revised_final.pdf, 2012.

Perrels, A.: Social economic benefits of enhanced weather services in Nepal – part of the Finnish Nepalese Project (FNEP), MFA intervention code: 79811801, Helsinki, 2011.

Perrels, A., Nurmi., V., and Nurmi, P.: Weather service chain analysis (WSCA) – An approach for appraisal of the social-economic benefits of improvements in weather services, SIRWEC Conference, Helsinki, 23–25 May 2012, paper 0081, http://www.sirwec2012.fi/Extended_Abstracts/002_Perrels.pdf, 2012.

World Bank: Weather and Climate Services in Europe and Central Asia, World Bank Working Paper No. 151, Washington DC, http://web.worldbank.org/WBSITE/EXTERNAL/COUNTRIES/ECAEXT/0,,contentMDK:22000203~pagePK:146736~piPK:146830~theSitePK:258599,00.html, 2008.

WMO: Socio-Economic Benefits of Meteorological and Hydrological Services and Information: Madrid Action Plan, World Meteorological Organisation, Geneva, http://www.wmo.int/pages/themes/wmoprod/documents/madrid07_ActionPlan_web_E.pdf, 2007.

WMO: Proceedings of the WMO Regional Association VI (Europe) Conference on Social and Economic Benefits of Weather, Climate and Water Services, Lucerne, Switzerland, 3–4 October 2012, PWS-23/ROE-1, ftp://ftp.wmo.int/Documents/PublicWeb/dra/eur/SEB_Conf_Oct2011/PWS_23_ROE-1_en.pdf, 2012.

WMO website – Public Weather Programmes: http://www.wmo.int/pages/prog/amp/pwsp/SocioEconomicEventsandReports.htm, last access: 2 May 2013.

Ye, Z., Strong, C., Fay, L., and Shi, X.: Cost Benefits of Weather Information for Winter Road Maintenance, Western Transportation Institute, Montana State University, USA, 2009.

15

Study of the MLB parameterisation for change in surface solar irradiance with sun zenith angle in clear sky

Z. Qu[1], **P. Blanc**[1], **M. Lefèvre**[1], **L. Wald**[1], and **A. Oumbe**[2]

[1]Centre for Energy and Processes, MINES ParisTech, BP207, 06904, Sophia Antipolis, France
[2]German Remote Sensing Data Center, German Aerospace Center (DLR),
Postfach 1116, 82234 Wessling, Germany

Abstract. The MLB parameterisation (Modified Lambert-Beer, Mueller et al., 2004) describes the change in SSI with sun zenith angle (SZA) in clear-sky conditions. It applies to the direct and global SSI as well as their spectral distribution. We assess its performances by comparing its results to the outputs of the radiative transfer model libRadtran and standard interpolation procedures. The standard two-point fitting MLB function performs very well at SZA between 0° and 60° and fairly bad from 60° to 89.9°. A parameterisation made of four MLBs for four intervals (0°, 60°), (60°, 75°), (75°, 85°) and (85°, 89.9°) is also tested. This piecewise MLB parameterisation exhibits satisfactory performances at any SZA and outperforms standard linear interpolation techniques. 95 % of errors in global SSI are less than 1 W m^{-2} for each band and less than 5 W m^{-2} for total irradiance.

1 Introduction

A new direct method, Heliosat-4, is currently developed by the MINES ParisTech and the German Aerospace Center (DLR), aiming at estimating surface downwelling solar irradiance (SSI). The clear-sky module, an important part of this method, is based on the radiative transfer model (RTM) libRadtran (Mayer et al., 2010) and benefits from advanced products derived from recent Earth Observation missions (Oumbe et al., 2009). However, running a RTM in an operational mode is computer resources- and time-consuming. Mueller et al. (2004, 2009) suggest the use of parameterisations, among which the Modified Lambert-Beer function (MLB) to reduce the number of runs of RTM for the different sun zenith angles (SZA). Instead of calculating SSI values for each SZA with RTM runs, we use the MLB to interpolate SSI at any SZA by using SSI calculated by RTM at only two SZAs.

It is crucial that the interpolated SSIs obtained by the MLB are very close to those output from the RTM. The MLB applies to the direct and global components of SSI as well as their spectral distribution. The use of the MLB function is increasing in scientific papers. However, to our knowledge, the performances of the MLB have not been fully reported, though a few results are presented in Mueller et al. (2004).

The novelty of this article is twofold: firstly to establish the performances of the MLB function by comparing its outcomes to those from libRadtran, and secondly, to propose an extended-MLB parameterisation to overcome the errors observed for large SZA. In addition, we have compared these performances to standard interpolation procedures in order to demonstrate the benefit of the MLB. It is assumed that libRadtran is delivering the actual values that should be reproduced.

2 The Modified Lambert-Beer parameterisation

Let $E_{\text{TOA},\lambda}(\theta_s)$ be the irradiance received by a horizontal plane at the top of atmosphere for the wavelength λ and the SZA θ_s. At altitude z_n, the direct spectral irradiance $E_\lambda(z_n, \theta_s)$ is a function of the optical depth $\tau_\lambda(z_n)$ and is given by the Lambert-Beer function (Liou, 1980):

$$E_\lambda(z_n, \theta_s) = E_{\text{TOA},\lambda}(\theta_s) \cdot \exp\left(\frac{-\tau_\lambda(z_n)}{\cos(\theta_s)}\right) \qquad (1)$$

The Modified Lambert-Beer function is based on Eq. (1) by adding a correction parameter $\alpha_\lambda(z_n)$:

$$E_\lambda^{\text{MLB}}(z_n, \theta_s) = E_{\text{TOA},\lambda}(\theta_s) \cdot \text{MLB}_\lambda(z_n, \theta_s) \qquad (2)$$

$$\text{MLB}_\lambda(z_n,\theta_s)=\exp\left(\frac{-\tau_\lambda(z_n)}{\cos(\theta_s)^{\alpha_\lambda(z_n)}}\right) \quad (3)$$

The two unknown parameters, $\alpha_\lambda(z_n)$ and $\tau_\lambda(z_n)$, are evaluated from two known irradiances $E_\lambda(z_n,\theta_1)$ and $E_\lambda(z_n,\theta_2)$. The fitting angles θ_1 and θ_2 are set here to $0°$ and $60°$ as in Mueller et al. (2004). Solving the linear system yields:

$$\alpha_\lambda(z_n) = \frac{\ln\left(-\ln\left(\frac{E_\lambda(z_n,\theta_1)}{E_{\text{TOA},\lambda}(\theta_1)}\right)\right)-\ln\left(-\ln\left(\frac{E_\lambda(z_n,\theta_2)}{E_{\text{TOA},\lambda}(\theta_2)}\right)\right)}{\ln(\cos(\theta_2))-\ln(\cos(\theta_1))} \quad (4)$$

$$\tau_\lambda(z_n) = \exp\left(\ln\left(-\ln\left(\frac{E_\lambda(z_n,\theta_1)}{E_{\text{TOA},\lambda}(\theta_1)}\right)\right)+\alpha_\lambda(z_n)\cdot\ln(\cos(\theta_1))\right) \quad (5)$$

The fitting parameters $\alpha_\lambda(z_n)$ and $\tau_\lambda(z_n)$ are calculated respectively for direct and global SSI, and for each spectral band. Then, the diffuse SSI is calculated by subtracting the direct SSI to the global SSI.

3 Assessing performances of the MLB

The principle of the assessment is the following. A large set of SSI values is created by runs of the libRadtran: these are the reference SSI. We use the MLB function to compute the SSI for the same conditions and we compare them to the reference. The differences are synthesized by the means of standard quantities, such as the bias, the root mean square error (RMSE) or the percentile P95 of the frequency distribution. This is done for the direct, diffuse and global SSI for each of the 32 spectral bands of Kato et al. (1999) and the total SSI.

However, it should be noted that when dealing with the measurements, the deviations between MLB and ground measurements could be higher than those between MLB and libRadtran.

We use a Monte-Carlo technique to randomly select 1000 sets within the 8D-space defined by the 8 most influent inputs to libRadtran: aerosol optical thickness, type, and Angstrom coefficient, total column water, total column ozone, altitude of the ground, ground albedo and atmospheric profile. For each 8-tuple, libRadtran is run for 5 above surface altitudes – from 0 to 2 km by step 0.5 km, – 22 SZAs – from 0 to 75° by step 5°, from 75° to 87.5° by step 2.5° and 89.9°. For a given 8-tuple and a given above surface altitude, the SSI values at the fitting angles are used to compute the fitting parameters of the MLB. Then, the MLB function is used to compute the SSI for the other SZAs, and these assessed SSIs are compared to the reference SSIs.

We have weighted the random selection of the 8-tuple parameters by considering the average distribution in optical properties of the atmosphere. The probability density functions for all parameters, except aerosol optical thickness and Angstrom coefficient, and total column ozone, obey the uniform law. We have selected the chi-square law for aerosol

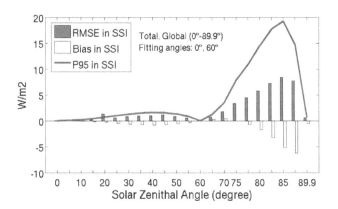

Figure 1. Errors in global total SSI $(0°, 60°)$.

optical thickness, the normal law for the Angstrom coefficient, and the beta law for total column ozone. The parameters of the laws have been empirically determined from the analyses of the observations made in the AERONET network for aerosol properties and from ozone products of the spatial missions OMI, GOME and SCHIAMACHY.

The results are illustrated in SSI or in clearness index KT which is defined as follow:

$$KT_\lambda(z_n,\theta_s) = E_\lambda(z_n,\theta_s)/E_{\text{TOA},\lambda}(\theta_s) \quad (6)$$

Using KT is a mean to show how good the method reproduces the effect of atmospheric contents on solar downwelling spectral irradiance, since the errors in SSI are expressed relatively to the irradiance at the top of atmosphere. Using KT instead of SSI is more convenient in several cases and both should be used. For example, the magnitude of the SSI changes from one Kato band to another and it is difficult to compare the deviation in SSI for different Kato bands. On the other side, errors in KT are commensurate and one can analyse and compare the errors for all cases more efficiently.

4 Results

4.1 MLB with fitting angles: 0° and 60°

Figure 1 displays the errors in SSI. It depicts the error in the total global SSI as a function of the SZA. For each SZA, the bias, RMSE, and P95 are computed using all differences in SSI in a 2D-space (1000 atmospheric conditions and 5 above surface altitudes). The errors are small for SZA less than 60° in full agreement with Mueller et al. (2004). 95 % of errors (P95) are less than 2 W m^{-2} below 60°, whereas the P95 may reach 20 W m^{-2} at 85°. As for errors in spectral SSI for the Kato bands (not shown), they follow the same trend: for any band, the P95 is less than 1 W m^{-2} below 60° and can be greater than 3 W m^{-2} between 75° and 85°.

Figure 2 displays the error in the total direct SSI as a function of the SZA. Globally, the performance of MLB for the direct SSI largely surpasses that for the global SSI.

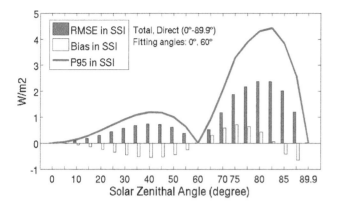

Figure 2. Errors in direct total SSI (0°, 60°).

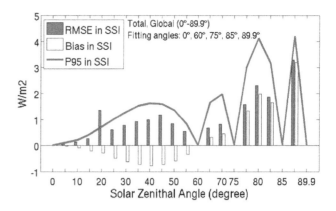

Figure 3. Errors for global total SSI (piecewise MLB).

P95 is less than $1.5\,\mathrm{W\,m^{-2}}$ below 60°, whereas the P95 may reach $4\,\mathrm{W\,m^{-2}}$ at 82.5°. Similar results are attained for spectral bands (not shown): for any band, the P95 is less than $0.1\,\mathrm{W\,m^{-2}}$ below 60° and attains its maximal value around $0.7\,\mathrm{W\,m^{-2}}$ between 80° and 85°.

According to Figs. 1 and 2, between 0° and 60°, the MLB performs very well for direct and global SSI, and thus for diffuse SSI. For greater SZAs, the error is much greater, especially for global SSI. Accordingly, the error in diffuse SSI is large. The MLB function tends to underestimate the diffuse SSI at large SZA, and therefore the global SSI, as demonstrated by the negative bias in Fig. 1. Similar conclusions are reached for each spectral band.

As discussed later and by Mueller et al. (2004), the usage of the modified Lambert-Beer function is physically motivated for direct irradiance; unsurprisingly, it does not perform very well for the diffuse irradiance. In addition, it is a fitting function and performs well within the range of SZA used for fitting and badly outside. Finally, the error in diffuse SSI or KT increases with ground albedo because the diffuse fraction increases with the ground albedo. As expected, there is no influence of the ground albedo on the direct component.

4.2 Piecewise MLB

In order to reduce errors for large SZA, we have studied performances of the MLB function with more-spaced fitting angles, such as (0°, 70°), (0°, 75°), or (0°, 80°). In all cases, the performances were very poor and inacceptable. We have also tested other couples of fitting angles: (60°, 75°) and (75°, 85°). Expectedly, we observed that the errors are always small for SZA comprised between these closely-spaced fitting angles, but increase rapidly outside these intervals.

We study now the parameterisation made of four MLBs whose parameters are assessed for four intervals (0°, 60°), (60°, 75°), (75°, 85°), and (85°, 89.9°). Now, we need 5 values of SSI for any situation. SSI at any SZA located in one of the four intervals is calculated by the MLB for this interval. Numerical errors may occur when solving Eqs. (4)

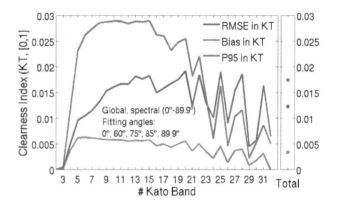

Figure 4. Errors for KTglo for Kato bands and total (piecewise MLB).

and (5) at 89.9°. To ensure convergence, the clearness index for the direct SSI KTdir is set to 1×10^{-15} when it is less than this threshold.

Figure 3 exhibits the error in global total SSI as a function of the SZA. Global SSI shows greater errors than direct SSI (not shown). The bias, RMSE, and P95 errors are small. The maximum of RMSE, bias and P95 for all SZA is respectively less than $4\,\mathrm{W\,m^{-2}}$, $4\,\mathrm{W\,m^{-2}}$ and $5\,\mathrm{W\,m^{-2}}$. We obtain most of the maximal errors in SSI when SZA equals to 87.5°.

Figure 4 displays the error in the clearness index of the global SSI KTglo for the interval [0°, 89.9°] for the Kato bands 3 (295 nm) to 32 (4298 nm) as well as the total. Errors are calculated in each band for all conditions. Errors are almost uniformly distributed for most of the Kato bands. The bias is very low, i.e., very low systematic error. The RMSE and P95 errors are small. The maxima of P95 are close to 0.030, reached for bands 8–15 (425–684 nm); they are always less than 4 % of the corresponding mean KTglo. As for total, P95 is equal to 0.017, i.e., less than 3 % of the mean of KTglo (0.68). Bias in SSI is negligible for each band and total, and P95 is less than $1\,\mathrm{W\,m^{-2}}$ for each band and $5\,\mathrm{W\,m^{-2}}$ for total. Errors in direct SSI are less than those for global SSI both in absolute and relative values.

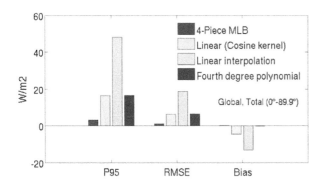

Figure 5. Comparison of errors for global total SSI (global performance of the four methods).

The clearness index is a convenient way to compare the results with different scales (large differences in SSI between Kato bands, and between Kato bands and total). In low SZA conditions, errors in SSI are generally small for Piecewise MLB (P95 < 5 W m^{-2}) as what we showed in the Fig. 3. However, as SZA increases, the horizontal irradiance at the top of atmosphere decreases and a small error in SSI could lead to a large error in clearness index. These large errors in clearness index in high SZA conditions then disturb the global statistics. For example, the percentile 95 often locates in high SZA cases like 82.5° where the error of 0.03 in clearness index is close to 5 W m^{-2} in global total SSI. Furthermore, the obtained RMSE in KT is less than 0.02 for each Kato band (Fig. 4), showing that deviations are generally much lower.

5   Comparisons with other techniques

We compared the performances of the piecewise MLB parameterisation with those attained by current interpolation techniques: linear interpolation, linear interpolation with cosine kernel between the same fitting angles: 0°, 60°, 75°, 85°, 89.9°. In addition, considering that we need five SZA to define the piecewise MLB, we have tested the performances of a fourth degree polynomial. However, this method performs fairly badly for the large interval between 0° and 60°. After multiple tests, we have selected five other fitting angles for the fourth degree polynomial: 0°, 20°, 45°, 70° and 89.9°. This set provides the best performances.

Figure 5 displays the performances of the four techniques for global total SSI for all conditions. This graph clearly demonstrates that the piecewise MLB outperforms the others. In terms of calculation time, the piecewise MLB is a bit slower than interpolation and has a significant advantage comparing to the fourth degree polynomial.

6   Conclusions

We have found that the MLB parameterisation (Mueller et al., 2004) is efficient for all wavelengths, provided the SZA

ranges between 0° and 60°. However, errors are large for SZA greater than 60°. The MLB function has a solid physical sense for the direct irradiance, and very good performances can be achieved. It is less physically sound for the diffuse irradiance and performances are lower for the diffuse and global irradiances.

After several tests, we have designed and validated a piecewise MLB parameterisation that offers very good performances for all wavelengths and all SZAs ranging from 0° to 89.9°. Errors for direct SSI are less than those for global SSI in absolute and relative values. The bias for global SSI is less than 0.2 W m^{-2} for each band and less than 4 W m^{-2} for total irradiance. 95% of errors in global SSI are less than 1 W m^{-2} for each band and less than 5 W m^{-2} for total irradiance. The piecewise MLB parameterisation outperforms current linear interpolation, cosine-based techniques and forth degree polynomial.

Acknowledgements. The research leading to these results has received funding from the European Union's Seventh Framework Programme (FP7/2007-2013) under Grant Agreement no. 218793 (MACC project).

Edited by: E. Batchvarova
Reviewed by: three anonymous referees

References

Kato, S., Ackerman, T., Mather, J., and Clothiaux, E.: The k-distribution method and correlated-k approximation for short-wave radiative transfer model, J. Quant. Spectrosc. Ra., 62, 109–121, 1999.

Liou, K. N.: An Introduction to Atmospheric Radiation, Academic Press, p. 182, 1980.

Mayer, B., Kylling, A., Emde, C., Buras, R., and Hamann, U.: libRadtran: library for radiative transfer calculations, Edition 1.0 for libRadtran version 1.5-beta, http://www.libradtran.org, 2 February 2010.

Mueller, R., Dagestad, K. F., Ineichen, P., Schroedter, M., Cros, S., Dumortier, D., Kuhlemann, R., Olseth, J. A., Piernavieja, G., Reise, C., Wald, L., and Heinnemann, D.: Rethinking satellite based solar irradiance modelling – The SOLIS clear sky module, Remote Sens. Environ., 91(2), 160–174, 2004.

Mueller, R. W., Matsoukas, C., Gratzki, A., Behr, H. D., and Hollmann, R.: The CM-SAF operational scheme for the satellite based retrieval of solar surface irradiance – A LUT based eigenvector hybrid approach, Remote Sens. Environ., 113, 1012–1024, 2009.

Oumbe, A., Blanc, Ph., Ranchin, T., Schroedter-Homscheidt, M., and Wald, L.: A new method for estimating solar energy resource, in: Proceedings of the ISRSE 33, Stresa, Italy, 4–9 May 2009, published by Joint Research Center, Ispra, Italy, USBKey, paper 773, 2009.

16

Formation of a large scale spot-like structure of the total deposition due to a powerful elevated finite source

O. Y. Skrynyk[1], R. I. Chernysh[1], and Y. Y. Hrytsyuk[2]

[1]Ukrainian Research Hydrometeorological Institute, Kyiv, Ukraine
[2]Ternopil National Pedagogical University, Ternopil, Ukraine

Abstract. The process of formation of a large scale spot-like structure of the cumulative deposition pattern due to a powerful elevated finite lifetime source is studied using numerical solutions of a model diffusion problem. A possible cause of spottiness at regional scale, namely the diurnal cycle of the vertical turbulent mixing intensity in the atmospheric boundary layer, is investigated. It is shown that the atmospheric boundary layer diurnal cycle can cause the appearance of the spot-like structure when the source lifetime, t_*, is much shorter than the diurnal period, T.

1 Introduction

A cumulative deposition pattern, which takes place under intense atmospheric emissions, has very distinct spot-like structure on different spatial scales. In the other words, a field of deposited materials is self-similar (multi-fractal distribution of deposited materials) (Randle et al., 2003).

The spot-like pollution structure at the regional scale (with linear size of $10^5 \div 10^6$ m or more) can be observed particularly well on a map that shows the distribution of the major pollutant, ^{137}Cs, on the territory of Ukraine after Chornobyl accident (see Fig. 1b). For comparison, in Fig. 1 we also show a map of ^{137}Cs distribution before the accident. Hereinafter by using the term spot-like structure at the regional scale we denote an existence of several local maxima in total deposition.

It is believed that the main reason for the formation of spots at regional scales is the convective cloud activity, i.e. "damp" aerosol scavenging with atmospheric precipitation ("washout") (see e.g. Clark and Smith, 1988; Puhakka, 1990). However precipitation was not observed when the southern radioactive trace was forming on the territory of Ukraine (Sedunov, 1989, Fig. 1). Taking into account the above mentioned fact we have hypothesized that along with other factors (precipitation, influence of the relief, surface inhomogeneity, etc.) a periodic change (e.g. diurnal cycle) of the vertical turbulent mixing intensity in the ABL can be a

reason causing the formation of a large scale spot-like deposition structure. Let us note that the importance of the diurnal variation of meteorological conditions in the process of formation of the earth surface pollution has been pointed out in many researches (see e.g. Baklanov and Sørensen, 2001; Brandt et al., 2002). However, the effect of the vertical turbulent mixing intensity in the ABL has not been studied in detail.

In our opinion, the hypothesis is quite reasonable and physically clear when the pollution is caused by an instantaneous elevated release (e.g. explosion, with subsequent transport and diffusion of the contamination cloud). For such a case the hypothesis has been confirmed by Voloshchuk et al. (2007). At the same time it is not clear whether the spottiness can be formed when the source is active for a limited time interval, as it occurred during the Chornobyl accident, and under what conditions this can happen.

It should be emphasized that the present work is not an attempt to reconstruct the deposition pattern that took place after the Chornobyl accident. The purpose of this work is to study the mentioned effect in the atmospheric diffusion using known and well approved methods, namely the K-theory. A simple but realistic schemes of parameterization of vertical structure and time evolution of the ABL have been used in the model diffusion problem. Note that the evolution of the diffusion processes considered in this work take place over space and time scales, where the use of the K-theory is justified.

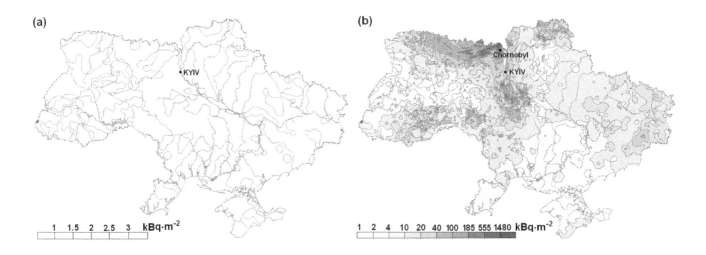

Figure 1. Maps of radioactive contamination of Ukraine by Cesium-137: **(a)** before-accident pollution (July 1985); **(b)** pollution after the Chornobyl accident (January 1998) (from Atlas of radioactive contamination of Ukraine (electronic version), 2002).

2 Parameterization of the atmospheric boundary layer

For the intensity of the vertical turbulent mixing, i.e. the value of the vertical eddy diffusivity, K_z, we will use the parameterization that was suggested in paper Mangia et al. (2002). In this parameterization the dependence of diffusivity on stability of ABL is introduced by means of the Obukhov-Monin length scale, L.

If $L < 0$

$$K_z = K_{zm} + K_{zc}, \quad K_{zm} = 0.3 z u_*(1 - z/h), \quad K_{zc} = \frac{0.24 z (z/h)^{1/3}}{q_{wc}^{4/3}} w_*,$$

$$w_* = u_* \left(-\frac{h}{\kappa L} \right)^{1/3},$$

$$q_{wc} = \begin{cases} 0.48, & z \leq 0.1h \\ 1.6 z/h (1 - e^{-4z/h} - 3 \times 10^{-4} e^{8z/h})^{-1}, & 0.1h \leq z \leq h \end{cases} \quad (1)$$

where K_{zm} is the mechanical part of the vertical diffusivity, K_{zc} is the convective part of the vertical diffusivity, h is the convective boundary layer height, w_* is the convective velocity scale, u_* is the friction velocity, κ is the von Kármán constant, q_{wc} is the stability function.

If $L > 0$

$$K_z = \frac{0.3 z u_*(1 - z/h)}{1 + 3.7 z/\Lambda}, \quad \Lambda = L(1 - z/h)^{(1.5\alpha_1 - \alpha_2)} \quad (2)$$

where α_1 and α_2 are empiric constants ($\alpha_1 = 2$, $\alpha_2 = 3$, Minnesota values).

The parameterization (1)–(2) belongs to a class of parabolic profiles which reproduce a real vertical profile of K_z, both in a day, and in a night ABL (Nieuwstadt and Van Dop, 1981).

The wind speed profile in the surface layer has been parameterized following the similarity theory of Monin-Obukhov (Monin and Yaglom, 1971)

$$u = \frac{u_*}{\kappa} [\ln(z/z_0) - \psi_m(z/L) + \psi_m(z_0/L)], \quad (3)$$

where $\psi_m(z/L)$ is the stability function. It represents the correction to the logarithmic profile due to non-neutral conditions ($\psi_m = 0$ in neutral conditions). The stability function is parameterized using the following relationships (Businger et al., 1971)

$$\psi_m = \begin{cases} 2\ln\left(\frac{1+\phi_m^{-1}}{2}\right) + \ln\left(\frac{1+\phi_m^{-2}}{2}\right) - 2\tan^{-1}\left(\phi_m^{-1}\right) + \frac{\pi}{2}, & L < 0 \\ -4.7 z/L, & L > 0 \end{cases}$$

$$\phi_m = \begin{cases} (1 - 16 z/L)^{-1/4}, & L < 0 \\ 1 + 4.7 z/L, & L > 0 \end{cases}$$

where ϕ_m is dimensionless wind profile.

We assume that above the surface layer the wind speed is constant: $u(z) = u(z_b)$, if $z > z_b$, where z_b is the surface layer height ($z_b \approx 5 \div 10\%$ of the ABL height).

The parameterisations (1)–(2), (3) give continuous values for K_z and u at all elevations and all stability conditions from unstable to stable, allowing us to treat dispersion in full diurnal cycle. As a rule, the unstable state prevails in the day time and the stable one at night (Stull, 1988).

To model diffusive processes over some days it is necessary to specify a diurnal cycle of stability. We will introduce the dependence of the Obukhov-Monin length scale on time as follows:

$$L^{-1} = B\cos\left(\frac{2\pi t}{T} + \omega_0\right), \quad (4)$$

where B is the amplitude of the diurnal change of L^{-1} (may depend on both the geographical coordinates and the time of the year), $T = 86400\,\text{s}$ is the full period, ω_0 is the initial phase, t is the time of the diffusion.

It should be emphasized that parameterization (4) is only the first approximation, which only qualitatively recreates the real diurnal change of the Obukhov-Monin length. Nevertheless, for the purpose of the present work, parameterization (4) can be considered satisfactory.

The parameterization (1)–(2) is 3-parametrical ($K_z = f(L, h, u_*)$). Parameters h and u_* also undergo a diurnal change. However, in our study we constrained ourselves by taking into account only the daily change of L, i.e. we assumed that h and u_* are constant (e.g. diurnal mean values can be used). The assumption of constant u_* is not very limiting, while the assumption of a constant h is a strong one and that has to be explained. At night h becomes small, the maximum value of K_z and its vertical position decrease. However at each elevation the day time value of K_z is larger than its night value. We believe that such behaviour of h and K_z would not change the results of our study qualitatively. Moreover, we believe that the diurnal change of h will intensify the spottiness effect (the spottiness will appear more distinctly). Because at night some amount of contamination, which happened to be above the height of the night ABL, will simply be transported and will not disperse.

3 Model diffusion problem

Let us consider a dispersion problem where the turbulent diffusion of gaseous and aerosol pollutants in the bottom part of the atmosphere is modeled. Assuming that the source of the pollutants is point-like and the pollutants are nonsettling, and orienting x-axis of the coordinate system in the direction of mean wind flow we obtain the following mathematical formulation of the problem:

$$\frac{\partial c}{\partial t} + u \frac{\partial c}{\partial x} = K_y \frac{\partial^2 c}{\partial y^2} + \frac{\partial}{\partial z} K_z \frac{\partial c}{\partial z}, \tag{5}$$

$$c = c(t, x, y, z), \quad t > 0, x > 0, |y| < \infty, z_0 < z < h,$$

$$K_y = \text{const.} \tag{6}$$

The initial condition is

$$c|_{t=0} = 0. \tag{7}$$

The boundary conditions are

$$cu|_{x=0} = QH(t)H(t_* - t)\delta(y)\delta(z - h_*),$$

$$Q = \text{const.}, \quad z_0 < h_* < h \tag{8}$$

$$c|_{|y| \to \infty} \to 0, \tag{9}$$

$$c|_{z=z_0} = 0, \ (z_0 \neq 0) \tag{10}$$

$$K_z \frac{\partial c}{\partial z}|_{z=h} = 0. \tag{11}$$

In expressions (5)–(11) we used the following notations: c is the concentration of pollutants; t, x, y, z are the time and space coordinates respectively; K_y is the crosswind eddy diffusivity; Q and h_* are the source strength and height; t_* is the source lifetime; $\delta(y)$ is the Dirac delta function; $H(t)$ is the Heavyside step function, defined as

$$H(t) = \begin{cases} 1, & t > 0 \\ 0, & t \leq 0 \end{cases}.$$

In Eq. (5) the wind speed, u, and the vertical eddy diffusivity, K_z, are defined by expressions (3) and (1)–(2) under the condition (4) respectively.

We assumed that K_y does not undergo a diurnal change ($K_y = \text{const.}$). The assumption is legitimate because it is known that the major contribution to the crosswind eddy diffusivity is made by the vertical wind shear. We also neglected the diffusion in the direction of the regular transport of pollutants ($K_x = 0$, where K_x is the eddy diffusivities along x direction), which allowed us to represent the source through the boundary condition (8). Such a trick is often used in modeling the turbulent diffusion in the presence of a continuous (or finite) source.

The upper boundary condition (11) along the vertical coordinate means that the transport of pollutants into the upper layers is absolutely forbidden. This does not represent a serious limitation for diffusion problems. Such a condition leads to somewhat overestimated intensities of the turbulent contaminant fluxes to the ground. As the lower boundary condition we use the condition of full absorption. Here, however, we do face a problem. The parabolic approximation of the vertical turbulent diffusivity (1)–(2) yields complete forbidding of the turbulent fluxes at $z = 0$. From purely mathematical standpoint it means that the boundedness condition is the only correct condition under such a parameterization. In other words, the parabolic parameterization of K_z is not compatible with the boundary condition of full (or partial) absorption. To resolve this problem we will lift the point where the lower boundary condition is given. The roughness height, z_0, can be used as a suitable value for this purpose (10).

Having obtained the solution of problem (5)–(11), (1)–(4) it is easy to find the total deposition to the ground, $c^*(x, y)$, using the following formula:

$$c^*(x, y) = \int_0^\infty K_z \frac{\partial c}{\partial z}|_{z=z_0} \, dt. \tag{12}$$

Here $c^*(x, y)$ is the principal unknown, which needs to be found.

It is practically impossible to obtain the analytic solutions of problem (5)–(11), (1)–(4). Thus, we must carry out a numerical simulation. For this purpose we used a finite-difference method. Using the method of sequential

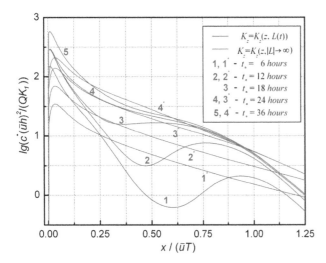

Figure 2. Horizontal profiles (at $y = 0$) of the normalized total deposition.

component-wise splitting (Marchuk, 1988) we obtained one-dimensional problems, to solve which we used the Crank-Nicholson method (Hoffman, 2001).

4   The analysis of the results – an empirical criterion for the appearance of a spot-like pollution structure

The results of the calculations of the total deposition based on the numerical simulations of diffusion problem considered above are shown in Fig. 2. There we present horizontal profiles of the total deposition $c^*(x,y)$ at $y = 0$ for several values of the source lifetime. The total deposition was normalized by $QK_1/(\bar{u}^2 h^2)$ where $\bar{u} = \frac{u(z_b)|_{L^{-1}=B} + u(z_b)|_{L^{-1}=-B}}{2}$, $K_1 = \frac{1}{h}\int_0^h K_z|_{L^{-1}=0} dz$, and is given in the logarithmic scale. The longitudinal coordinate x is normalized by the distance, which a unit volume of air passes for a period $T = 24$ h due to the regular transport ($\bar{u}T$).

The results of the total deposition calculations when a diurnal change of stability was neglected in the diffusion problem (i.e. when $|L| \to \infty$, that corresponds to neutral conditions) are placed in Fig. 2 as well. The latter has been done in order to show that the occurrence of more than one maxima of the total deposition is due to the periodic change of K_z only.

In all of the cases the actual values of the input parameters for the diffusion problem were kept the same. To obtain numerical solutions of diffusion problem we used the following values of parameters: $B = 0.01$ m^{-1}, according to paper Irwin (1979) it means that the stability in the ABL changes from Pasquill-Gifford stability class C to stability class E; $\omega_0 = \frac{\pi}{2}$, it corresponds to a situation when releases begin approxi-

mately at morning time; $z_0 = 0.1$ m; $z_b = 50$ m; $h_* = 500$ m; $h = 1000$ m; $Q = 1$ kg s^{-1}; $u_* = 0.35$ m s^{-1}; $\bar{u} = 6.12$ m s^{-1}; $K_y = 300$ m^2 s^{-1}; $K_1 = 17.5$ m^2 s^{-1}.

Based on the analysis of the obtained results we reached the conclusion that the effect of spottiness takes place not only in the presence of an instantaneous source, but also in the presence of a source with a finite lifetime. The spot-like pollution structure of the cumulative deposition can be clearly seen in Fig. 2 at relatively small values of t_* (function $c^*(x,0)$ has more than one maximum). When the lifetime of the source gets longer, the accumulation of the pollutants at the local minima of $c^*(x,0)$ becomes more intensive. This leads to some smoothing of the pollutant field. As t_* increases, all local maxima, except for the first one, disappear and function $c^*(x,0)$ becomes monotonically decreasing in the interval $x > x_{\max 1}$, where $x_{\max 1}$ is the position of the first maximum. At $t_* = T, 2T, 3T \dots$ the function becomes strictly monotone, while at $t_* \in (Ti, T(i+1))$, where $i \in N$, the strict monotony might not hold. It is difficult to explain such behavior of the total deposition. A possible reason is following: when the lifetime of the source gets longer, every location (x points) experiences different K_z value due to the different stability conditions (L value) varying during the daily cycle, while for very short puffs only a limited range of L values are "seen" when the puff travels over a specific location.

The existence of a certain criterion for the spottiness can be justified relatively easy. On the one hand, it is obvious that the periodic change of stability in the ABL will result in several local maxima of total concentration deposition when the instantaneous source acts. Qualitatively this process can be pictured as follows: the wind transports the cloud of contaminants and during this transportation the vertical turbulent fluxes of contaminants to the ground are "switched on" in the daytime and "switched off" at night. On the other hand, it is also obvious that the spottiness will not appear when $t_* \to \infty$ because of the effect of averaging of many puffs that form the plume. Thus, a finite source lifetime for which the spot-like pollution structure vanishes has to exist.

We carried out numeric experiments with different values of input model parameters, like B and ω_0 ($\omega_0 \in [0; 2\pi]$, $B \in [0.005; 0.05]$). The value of these parameters influences the maxima of the total deposition and their positions (x coordinates). But under any value of B and ω_0 the spot-like structure was observed.

The conclusions presented above are obtained based on the results of numerous computer simulations, i.e. they are empirical. It is very hard to prove them in a rigorous mathematical sense.

From the analysis of the results of the numerical simulations it follows that the effect of spottiness takes place when condition $t_*/T < 1$ is satisfied. This relation can be considered as an empirically established criterion for the appearance of a spot-like pollution structure.

5 Conclusions

Empirical data and experiments, which can help to understand the effect of diurnal change in the vertical turbulent mixing intensity on the surface contamination pattern, are very scarce. With the exception of deposition patterns generated by Chornobyl accident release, we do not know of any other relevant experimental data. Thus, at present mathematical modelling (numerical simulation) remains the only way to verify the hypothesis that spottiness may be caused by the diurnal cycle of the vertical turbulent mixing intensity in the ABL.

In this work we have studied the process of formation of the total dry deposition to the ground in the presence of a finite lifetime point source. From the analysis of the results it follows that the effect of large scale spottiness due to the diurnal cycle of the vertical turbulent mixing intensity in the ABL takes place when certain conditions are satisfied. We have formulated an empirical criterion of the appearance of a spot-like pollution structure, which is determined by the ratio of the source lifetime and the characteristic time of the periodic changes in the intensity of the vertical turbulent mixing.

Acknowledgement. Authors wish to thank anonymous reviewers for valuable comments and suggestions.

Edited by: D. Giaiotti
Reviewed by: two anonymous referees

References

Baklanov, A. and Sørensen, J. H.: Parameterisation of radionuclide deposition in atmospheric long-range transport modelling, Phys. Chem. Earth (B), 26(10), 787–799, 2001.

Brandt, J., Christensen, J. H., and Frohn, L. M.: Modelling transport and deposition of caesium and iodine from the Chernobyl accident using the DREAM model, Atmos. Chem. Phys., 2, 397–417, 2002, http://www.atmos-chem-phys.net/2/397/2002/.

Businger, J. A., Wyngaard, J. C., Izumi, Y., and Bradley, E. F.: Flux-profile relationship in the atmospheric surface layer, J. Atmos. Sci., 28, 181–189, 1971.

Clark, M. J. and Smith, F. B.: Wet and dry deposition of Chernobyl releases, Nature, 332, 245–249, 1988.

Atlas of radioactive contamination of Ukraine (electronic version), Kyiv, Ministry on Emergency Situations of Ukraine, 2002.

Irwin, J. S.: A theoretical variation of the wind profile power-law exponent as a function of surface roughness and stability, Atmos. Environ., 13, 191–194, 1979.

Hoffman, J. D.: Numerical methods for engineers and scientists, Basel, New York, 2001.

Mangia, C., Moreira, D. M., Schipa, I., Degrazia, G. A., Tirabassi, T., and Rizza, U.: Evaluation of a new eddy diffusivity parameterisation from turbulent Eulerian spectra in different stability conditions, Atmos. Environ., 36, 67–76, 2002.

Marchuk, G. I.: Metody raschepleniya (Methods of splitting), Nauka, Moscow, 1988.

Monin, A. S. and Yaglom, A. M.: Statistical fluid mechanics, MIT Press, Cambridge, 1971.

Nieuwstadt, F. T. M. and Van Dop, H. (Eds.): Atmospheric turbulence and air pollution modeling. A course held in the Hague, 21–25 September, Reidel Publishing Company, 1981.

Puhakka, T.: Meteorological factor influencing the radioactive deposition in Finland after the Chernobyl accident, J. Appl. Meteor., 29, 813–829, 1990.

Randle, K., Sokhi, R. S., and Cooper, J. R.: Radioactive Releases in the Environment: Impact And Assessment, John Wiley & Sons, Chichester, 473 pp., 2003.

Sedunov, Yu. S., Borzilov, V. A., Klepikova, N. V., Chernokozhin, E. V., and Troyanova, N. I.: Physicomathematical modeling of the regional transport of radioactive pollutants in the atmosphere in consequence of the Chernobyl accident, Russ. Meteorol. Gidrol., 9, 5–10, 1989.

Stull, R. B.: An introduction to boundary layer meteorology, Kluwer Academic Publishers, Dordrecht, 666 pp., 1988.

Voloshchuk, V. M., Skrynyk, O. Y., and Hrytsyuk, Y. Y.: Formation mechanism of a large-scale spot-like pollution structure of the underlying surface under intense aerosol injections into the atmospheric boundary layer, Reports of the National Academy of Sciences of Ukraine, 4, 115–120, 2007.

17

Climatology of persistent deep stable layers in Utah's Salt Lake Valley, USA

S. Zhong[1], **X. Xu**[1], **X. Bian**[2], **and W. Lu**[1]

[1]Department of Geography, Michigan State University, East Lansing, Michigan USA
[2]USDA Forest Service Northern Research Station, East Lansing, Michigan, USA

Abstract. The characteristics of winter season persistent deep stable layers (PDSLs) over Utah's Salt Lake Valley are examined using 30-year twice daily rawinsonde soundings. The results highlight the basic climatological characteristics of the PDSLs, including the strengths of the inversion, the frequency of the occurrence, and the duration of the events. The data analyses also reveal linear trend, interannual variability, as well as the relationship between the interannual variability of PDSLs and the variability of large-scale circulations. Finally, the study investigates the large-scale atmosphere conditions accompanying the formation and destruction of the PDSL episodes.

1 Introduction

Persistent deep stable layers (PDSL) are frequent wintertime phenomena in basins and valleys of the Intermountain West of the United States (Wolyn and Mckee, 1989; Reeve and Stensrud, 2009). The strong static stability decouples the air within the basin/valley from layers aloft, which often leads to stagnation and limits atmospheric dispersion (Vosper and Brown, 2008; Smith et al., 2010). Cold and sometimes moist air, being trapped in basins or valleys by PDSL, favors the formation of fog, drizzle, or freezing rain, which can be hazardous for ground transportation as well as aviation. Accurate forecasting of the formation and removal of these PDSLs has proven to be one of the many challenges of mountain weather forecasting (Smith et al., 1997; Zängl, 2005).

In this study, we perform climatological analyses of winter season PDSLs over the Salt Lake Valley (SLV) in Utah, USA. The objectives are to determine the basic climate characteristics of PDSLs, understand their interannual variability and trend, and understand the synoptic conditions that accompany the buildup and breakup of PDSLs.

2 Site, data, and method

The SLV (Fig. 1) is bordered by high terrain on three sides: the Oquirrh Mountains to the west, the Wasatch Range to the east, and the Traverse Range to the south. The highest peak along the Wasatch Range has an elevation of 3300 m above mean sea level (MSL), approximately 2000 m above the valley floor. The valley opens up to the north with the Great Salt Lake to the northwest.

The analysis employed winter-season (defined here as the beginning of November through the end of February), twice-daily rawinsonde soundings launched from the Salt Lake City International Airport in the north central part of the SLV (Fig. 1) for the period of 1979 to 2009.

For this study, a PDSL event is defined as $\frac{\partial \theta}{\partial z}\big|_{z_2 - z_1} > 0.0033\,°\text{C}\,\text{m}^{-1}$ (Z_1 and Z_2 are 850-hPa and 700-hPa geopotential heights) is satisfied by at least three consecutive soundings. This is equivalent to the condition that the temperature lapse rate between 850 and 700 hPa is less than the moist adiabatic lapse rate; in other words, the atmosphere is absolutely stable. All PDSL events are further classified into three categories based on the value of the potential temperature gradient: a strong PDSL event if

$$\frac{\partial \theta}{\partial z}\bigg|_{z_2 - a_1} > 0.017\,°\text{C}\,\text{m}^{-1};$$

a weak event if

$$0.0033\,°\text{C}\,\text{m}^{-1} < \frac{\partial \theta}{\partial z}\bigg|_{z_2 - a_1} \leq 0.009\,°\text{C}\,\text{m}^{-1};$$

and a moderate event, otherwise.

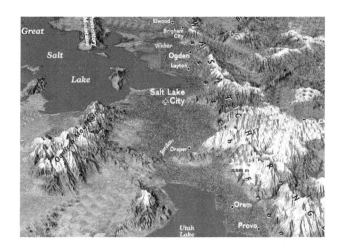

Figure 1. The Salt Lake Valley of Utah, USA (NASA Satellite Image; Courtesy of Frank Ludwig).

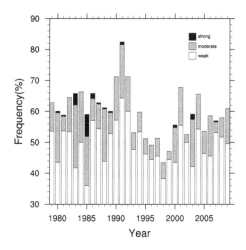

Figure 2. Winter season mean frequency of PDSL occurrence for the period of 1979–2009.

3   Results and discussions

3.1   Frequency of occurrence

PDSL is a frequent winter season phenomenon over the SLV. As shown in Fig. 2, the winter season mean PDSL frequency of occurrence ranges from 44% in 1998 to 83% in 1991, with considerable interannual variability. Weak PDSLs are much more frequent than the stronger events except for four winters when the frequency of moderate PDSLs slightly exceeds that of weak ones. Strong PDSLs are rare events. An analysis of the relationship between the interannual variability of PDSL frequency and large-scale circulation indices (Fig. 3) revealed a weak correlation ($r = 0.4365$, $P < 0.05$) between the interannual variation of the frequency of the weak PDSLs and the MEI index, a composite El Niño-Southern Oscillation (ENSO) index with positive (negative) values representing the warm (cold) ENSO phase. No clear relation-

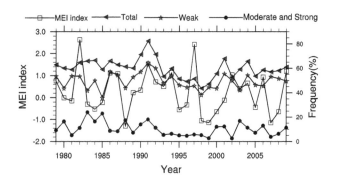

Figure 3. The frequency of PDSL events and the MEI index.

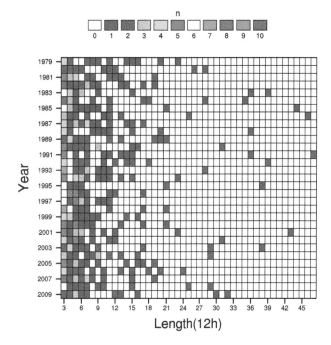

Figure 4. The length of PDSLs and the frequency of their occurrence. The color represents the number of times a PDSL of given length occurs in the year.

ship, however, is detected between MEI and the interannual variability of the frequency of the moderate or strong PDSL events. The linear trend analysis (not shown) indicates no trend for the weak PDSL events, but a slight downward trend for the moderate and strong events.

3.2   PDSL length

How long do PDSL events last? As shown in Fig. 4, a PDSL may last from 1.5 day (3 consecutive 12 h soundings) to a little over 3 weeks (42 consecutive 12 h soundings). As expected, the longer PDSL events occur much less frequently than shorter ones. With a few exceptions, PDSL events lasting longer than 1 week rarely happen twice in a single year. PDSL events shorter than 1 week account for 86% of the total

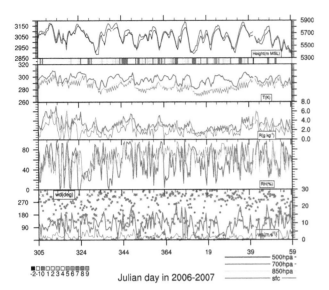

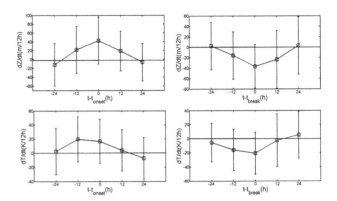

Figure 6. The mean time rate of change and the standard deviation of 500-hPa geopotential heights (top two) and 700-hPa temperature (bottom two) 24 h prior to and after the onset and breakup of a PDSL.

Figure 5. Time series of geopotential height (top), potential temperature (2nd from the top), mixing ratio (middle), relative humidity (2nd from the bottom), and wind speed/direction (bottom). The color bar indicates inversion strength (−2 missing data, and −1 no inversion).

frequency, with the rest of 14% for those lasting longer than 1 week. While it is possible for a PDSL event to last longer than 2 weeks, they are rare events with only 3% frequency of occurrence.

3.3 Synoptic conditions accompanying the buildup and breakup of PDSL

Time series of geopotential height, potential temperature, moisture, and wind at various atmospheric levels are examined for all 30 winters to understand synoptic conditions accompanying PDSL events and an example is shown in Fig. 5 for the winter season of 2006–2007. The buildup (breakup) of PDSLs is usually accompanied by rising (falling) 700-hPa and 500-hPa geopotential heights associated with an approaching ridge (trough) and by 700-hPa warming (cooling). To further illustrate the role of synoptic advection, the mean rate of change of 500-hPa geopotential height and 700-hPa temperature prior to and after the onset and breakup of all PDSL events is examined and the results are shown in Fig. 6. The positive (negative) 500 hPa height advection typically starts 24 h prior to the buildup (breakup) and the rate of height increase (decrease) peaks around the time of the PDSL onset (breakup). Warm (cold) advection also tends to begin 24 h before the buildup (breakup). The rate of 700 hPa warming peaks around 12 h before the onset, and the maximum rate of cooling occurs around the time of breakup.

To understand the influence of wind aloft on the development of PDSLs, the 700-hPa wind speed and direction with/without PDSLs are compared and the results (not

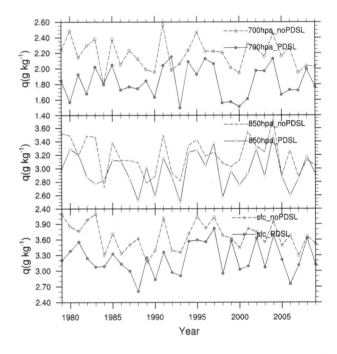

Figure 7. Specific humidity averaged over all PDSL days and all non PDSL days, respectively.

shown) reveal a 10–20% wind speed reduction during PDSL events, but no systematic difference in wind direction. A comparison of surface wind with/without PDSLs reveals a considerably lower wind speed in the valley with the presence of PDSLs.

3.4 PDSL and humidity

PDSLs are characterized by consistently lower specific humidity both within and above the valley (Fig. 7), indicating generally drier atmospheric conditions accompanying PDSLs. The relative humidity is always lower at 700 hPa,

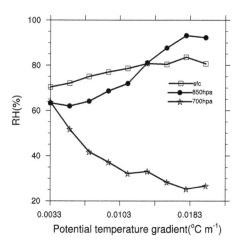

Figure 8. Changes of relative humidity as a function of potential temperature gradient.

a result of both warming and drying, and the stronger the inversion, the lower the relative humidity (Fig. 8). Within the valley, however, the relative humidity is always higher during PDSLs due mainly to colder temperature, and the stronger the inversion, the higher the relatively humidity (Fig. 8). At very strong stability, a further increase in the inversion strength does not lead to a continuing increase or decrease in the relative humidity

4 Future work

The analysis performed here is being extended to other basins and valleys in Western US. The results from different basins/valleys will be compared to determine the spatial variability of the PDSL characteristics.

Acknowledgement. This material is based upon work supported by the National Science Foundation under Grant No. 0938401. Any opinions, findings, and conclusions or recommendations expressed in this material are those of the author(s) and do not necessarily reflect the views of the National Science Foundation. The authors greatly appreciate the comments from the two anonymous referees and from C. J. Ander, J. Horel, and C. D. Whiteman.

Edited by: M. Sprenger
Reviewed by: two anonymous referees

References

Reeve, H. D. and Stensrud, D. J.: Synoptic-scale flow and valley cold pool evolution in the western United States, Wea. Forecasting, 24, 1625–1643, 2009.

Smith, R., Paegle, J., Clark, T., Cotton, W., Durran, D., Forbes, G., Marwitz, J., Mass, C., McGinley, J., Pan, H.-L., and Ralph, M.: Local and remote effects of mountains on weather: Research needs and opportunities, B. Am. Meteorol. Soc., 78, 877–892, 1997.

Smith, S. A., Brown, A. R., Vosper, S. B., Murkin, P. A., and Veal, A. T.: Observations and simulations of cold air pooling in valleys, Bound.-Lay. Meteorol., 134, 85–108, 2010.

Vosper, S. B. and Brown, A. R.: Numerical simulations of sheltering in valleys: The formation of nighttime cold-air pools, Bound.-Lay. Meteorol., 127, 429–448, 2008.

Wolyn, P. G. and Mckee, T. B.: Deep stable layers in the intermountain western United State, Mon. Weather Rev., 117, 461–472, 1989.

Zängl, G.: Dynamical aspects of wintertime cold-air pools in an Alpine valley system, Mon. Weather Rev., 133, 2721–2740, 2005.

18

Large-eddy simulation of plume dispersion within regular arrays of cubic buildings

H. Nakayama[1], K. Jurcakova[2], and H. Nagai[1]

[1]Japan Atomic Energy Agency, Ibaraki, Japan
[2]Institute of Thermomechanics, Academy of Sciences of the Czech Republic, Prague, Czech Republic

Abstract. There is a potential problem that hazardous and flammable materials are accidentally or intentionally released within populated urban areas. For the assessment of human health hazard from toxic substances, the existence of high concentration peaks in a plume should be considered. For the safety analysis of flammable gas, certain critical threshold levels should be evaluated. Therefore, in such a situation, not only average levels but also instantaneous magnitudes of concentration should be accurately predicted. In this study, we perform Large-Eddy Simulation (LES) of plume dispersion within regular arrays of cubic buildings with large obstacle densities and investigate the influence of the building arrangement on the characteristics of mean and fluctuating concentrations.

1 Introduction

An accurate analysis of plume dispersion is important for emergency responses against accidental or intentional release of hazardous and flammable materials within populated urban areas. For the assessment of human health hazard or the safety analysis of the hazardous gas, not only mean but also fluctuating concentrations should be estimated, considering the effects of individual buildings. Therefore, the dispersion characteristics of a plume through obstacle arrays have been examined mainly by field and wind tunnel experiments. For example, Davidson et al. (1996) investigated the influence of building arrays on plume dispersion by wind tunnel experiments. Bezpalcova and Ohba (2008) conducted wind tunnel experiments of plume dispersion within various building arrays and investigated the effects of the building arrangement and obstacle density on the characteristics of mean and root mean square (RMS) concentrations.

In this study, we perform numerical simulations of plume dispersion within a regular array of cubic buildings as idealized urban canopy by Large-Eddy Simulation (LES) that can give detailed information on turbulent flow and concentration fields. The objective of this study is to perform LES of plume dispersion within building arrays with large obstacle densities, which corresponds to densely built-up urban areas

and investigate the distribution patterns of concentrations and the characteristics of peak concentration within the building array.

2 Numerical model

The basic equations for the LES model are the spatially filtered continuity equation, Navier-Stokes equation and the transport equation for concentration. The subgrid-scale (SGS) Reynolds stress is parameterized by using the standard Smagorinsky model (Smagorinsky, 1963), where the Smagorinsky constant is set to 0.1 for estimating the eddy viscosity (Murakami et al., 1987). The subgrid-scale scalar flux is also parameterized by an eddy viscosity model and the turbulent Schmidt number is set to 0.5.

The coupling algorithm of the velocity and pressure fields is based on the Marker and Cell (MAC) method (Harlow and Welch, 1965) with the second-order Adams-Bashforth scheme for time integration. The Poisson equation is solved by the Successive Over-Relaxation (SOR) method which is an iterative method for solving a Poisson equation for pressure. For the spatial discretization in the governing equation of the flow field, a second-order accurate central difference is used. For the dispersion field, Cubic Interpolated Pseudo-particle (CIP) method proposed by Takewaki et al. (1985) is used for the advection term. CIP is a very stable scheme that can solve generalized hyperbolic equations in space. For diffusion term, a second-order accurate central difference method is used. The time step interval $\Delta t U_\infty / H$ is 0.005 (Δt: time step). The maximum CFL (Courant-Friedrich-Levy) number is about 0.15.

3 Test simulations

3.1 Wind tunnel experiments for evaluating the model performance

The experiments were carried out by Bezpalcova and Ohba (2008) in the Boundary Layer Wind Tunnel at Wind Engineering Center of Tokyo Polytechnic University, Japan. The experimental set-up consists of buildings with dimensions: 70 mm (width), 70 mm (length), and 70 mm (height). In this paper, obstacle density λ_f is defined as the ratio of the total floor projection area of buildings to the plan area of the study site. Buildings are arranged in the regularly square array with $\lambda_f = 0.25$ and 0.33. There are 18×9 and 20×9 building arrays with $\lambda_f = 0.25$ and 0.33, respectively. The ground-level point source is located at the center just behind a building of the 8th row and the 5th column, and the 9th row and the 5th column of the arrays in cases of $\lambda_f = 0.25$ and 0.33, respectively. Here, the rows are numbered in increasing order in the streamwise direction from the leading edge of the array and the columns are numbered in increasing order in the spanwise direction. In their experiment, the lower part of the neutral atmospheric boundary layer is simulated by vortex generators set up at the wind tunnel section and roughness blocks as shown in Fig. 1. The scale of the modeled boundary layer is 1:400, i.e. the boundary layer height corresponds to 120 m in the full scale. The mean wind velocity vertical profile of approach flow can be approximated by a power law exponent of 0.25. Wind velocity was measured by Thermoanemometry using a split-fibre probe. The uncertainties of flow measurement were 5% for both mean and RMS quantities. Concentration is measured using a fast-response flame ionization detector. The uncertainties of concentration measurement were 9% and 17% for mean and RMS quantities, respectively. In this wind tunnel experiment, the building Reynolds numbers based on the cubical building height and wind speed at the building height is about 14 000.

In this study, to evaluate the model performance, we compare our LES results with these wind tunnel experimental data.

3.2 Computational settings

Figure 2 shows a schematic illustration of the numerical model. Two computational domains are set up: The main region for a simulation of plume dispersion within a building array and the driver region for generating a spatially-developing turbulent boundary layer flow. First, a thick turbulent boundary layer flow is generated by incorporating the inflow turbulence generation method of Kataoka and Mizuno (2002) into an upstream part of the driver region and, then, a

Figure 1. The wind tunnel set-up.

wind flow with strong turbulent fluctuations is produced by a tripping fence and roughness blocks placed at the downstream of the recycle station. The fluctuating part of the velocity at the recycle station is recycled and added to the specified mean wind velocity as shown in Fig. 2a. This unsteady wind flow is imposed at the inlet of the main region at each time step and calculations of turbulent flow and plume dispersion within a building array are performed as shown in Fig. 2b.

In the driver region, the Sommerfeld radiation condition (Gresho, 1992) is applied at the exit, a free-slip condition for streamwise and spanwise velocity components is imposed and vertical velocity component is 0 at the top. A periodic condition is imposed at the side and a non-slip condition for each velocity component is imposed at the ground surface. The size and the number of grid points for the driver region is $13.8\delta_L \times 3.8\delta_L \times 5.0\delta_L$ (δ_L: the scale of the modeled boundary layer) and $460 \times 250 \times 100$ in streamwise, spanwise and vertical directions, respectively. A tripping fence and each roughness block set up in the driver region are resolved by $7 \times 250 \times 24$ and $3 \times 6 \times 12$ grids in streamwise, spanwise and vertical directions, respectively. The Van Driest damping function (Van Driest, 1956) is incorporated to account for near-wall effects and the resolution of a grid above the ground surface is set to 2.0. Building effects are represented by the feedback forcing method proposed by Goldstein et al. (1993). The main idea of this method is to apply the external force inside the body.

In the main region, there are 25×8 and 28×9 obstacle arrays with $\lambda_f = 0.25$ and 0.33, respectively. The ground-level point source is located just behind a building of the 8th row and the 4th column, and the 9th row and the 5th column of

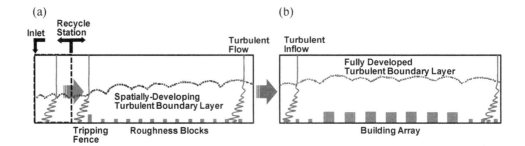

Figure 2. Schematic diagram of the numerical model. (**a**) Driver region for generating boundary layer flow. (**b**) Main region for plume dispersion within a building array.

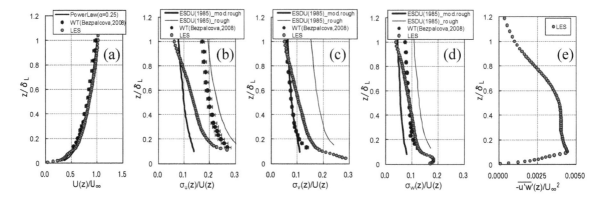

Figure 3. Turbulence characteristics of approach flow. (**a**) Mean wind velocity. (**b**) Streamwise turbulence intensity. (**c**) Horizontal turbulence intensity. (**d**) Vertical turbulence intensity. (**e**) Reynolds stress.

the arrays in cases of $\lambda_f = 0.25$ and 0.33, respectively. Each building of the array is resolved by $16 \times 16 \times 24$ grids in the streamwise, spanwise and vertical directions, respectively. At the inlet of the main region, the inflow turbulence data obtained near the exit of the driver region is imposed. The other boundary conditions in a flow field are the same as those in the driver region but the damping function to account for near-wall effects is not incorporated. In a concentration field, zero gradient is imposed at all the boundaries (Shi et al., 2008). Assuming that the location of a plume source point in the wind tunnel experiment is $x/H = 0.0$, $y/H = 0.0$ and $z/H = 0.0$ (H: a building height), that of a plume source point in this LES model is $x/H = 0.0$, $y/H = -0.03$ and $z/H = 0.0$. Because the number of grid points for individual cubic building is even number, the plume source position in the y/H coordinate is slightly different from that in the experimental condition. The size and the number of grid points for the main region are $18.0\delta_L \times 3.8\delta_L \times 5.0\delta_L$ and $1000 \times 250 \times 100$ in streamwise, spanwise and vertical directions, respectively. The lengths of the domain in front of the first row and behind the last row are both $3.0\delta_L$. The grid resolution above the ground surface is the same as the one in the driver region but the Van Driest damping function is not used in the main region.

The length of the simulation run to calculate the time averaged values of velocity and concentration TU_∞/H (T: averaging time) is 500. The length of the simulation run before releasing the scalar is TU_∞/H is 250. In the present LESs, the building Reynolds number is almost 5000.

4 Results

4.1 Approach flow

Figure 3 compares the LES results of turbulence characteristics of approach flow with the wind tunnel experimental data of Bezpalcova and Ohba (2008) and the recommended data of **E**ngineering **S**cience **D**ata **U**nit 85020 (ESDU 85020, 1985). ESDU 85020 provides comprehensive turbulence characteristics of neutrally stratified atmospheric boundary layer based on independent experimental measurements ranging from the ground surface to 300 m. ESDU 85020 recommends vertical profiles of turbulence intensities for each wind component and their relationship in dependence on surface roughness. δ_L corresponds to $4.0H$ and is assumed to be 120 m in the full scale condition. The experimental data are shown with the error bars described in Sect. 3.1. The profile of the mean wind velocity of LES is found to fit the experimental profile of 0.25 power law. LES

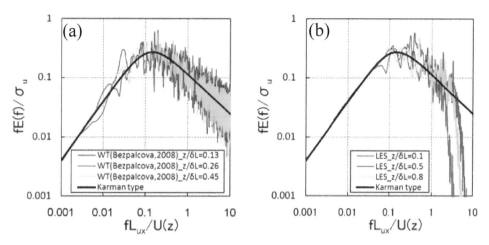

Figure 4. Power spectrum of approach flow obtained from (a) wind tunnel experiment and (b) LES.

Figure 5. Instantaneous plume dispersion field in case of $\lambda_f = 0.25$. The yellow areas on iso-surface indicate 0.01% of initial concentration. (a) $t* = 15.0$, (b) $t* = 54.0$, (c) $t* = 117.0$ after the plume release.

approach flow turbulence intensities are copying shape of vertical profiles recommended by ESDU 85020 very well between the recommended data for moderate rough and rough surfaces up to $0.8\delta_L$. The experimental data agree with LES and ESDU data only for vertical component, experimental streamwise and spanwise turbulence intensities are slightly overpredicted and underpredicted, respectively. The vertical profile of Reynolds stress of LES shows a constant profile in the range $0.1 < z/\delta_L < 0.5$. According to the review paper of Counihan (1975), it is shown that the average height of the constant shear stress layer is 100 m. The LES data lies within this range shown by Counihan (1975). Figure 4 shows the power spectrum of the approach flow of (a) the experiment and (b) LES. f, $E(f)$ and L_{ux} indicate the frequency, the longitudinal velocity spectra and the integral length scale, respectively. The each power spectra obtained by wind tunnel experiment is consistent with the Karman type. Although LES power spectra rapidly decrease in higher frequency side $fE(f)/\sigma_u > 2$, the LES data are found to show good agreement with the Karman type except the high frequency side.

The LES approach flow corresponds to a neutral atmospheric boundary layer based on comparison with the ESDU 85020 recommended data. Although some of the turbulence characteristics by LES are quantitatively different from those by the experiment, they both reasonable well model the neutral boundary layer above rough surface and can be compared taking in account their differences.

4.2 Dispersion characteristics

Figure 5 shows instantaneous plume dispersion fields in case of $\lambda_f = 0.25$ at times $t* (= tU_\infty/H) = 15.0$, 54.0 and 117.0 after the plume release. The yellow areas on iso-surface indicate 0.01% of initial concentration. It shows that a portion of the plume is moved upwards by the rising airflow behind the upstream building at first, and then the plume is transported in the streamwise direction with being entrained into the street canyon normal to the wind direction. After enough time passing, the plume is found to be transported and dispersed within and above a building array.

Figures 6 and 7 compare the LES results with the wind tunnel experimental data (Bezpalcova and Ohba, 2008) of the spanwise profiles of mean (C_{ave}) and RMS (C_{RMS}) concentrations at a height of $0.29H$ at the 4th and 5th row behind the source in $\lambda_f = 0.25$ and 0.33, respectively. The mean and RMS concentrations are normalized by wind velocity at the building height (U_H), the building height and the source strength (Q). The experimental data are shown with the error bars. In both cases of $\lambda_f = 0.25$ and 0.33, the spanwise spread of the plume of the wind tunnel experiments by Bezpalcova and Ohba (2008) is enhanced by the influence of buildings and high concentration region is formed in the range $-1.0 < y/H < 1.0$. The mean concentration decreases towards the plume edge. Although LES data overpredict in both cases slightly, the tendency such as the formation of the

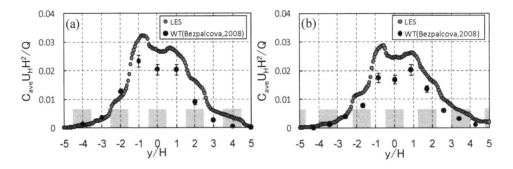

Figure 6. Spanwise profiles of mean concentration at a height of $0.29H$. **(a)** at the 4th row behind source location in the case of $\lambda_f = 0.25$. **(b)** at the 5th row behind source location in the case of $\lambda_f = 0.33$.

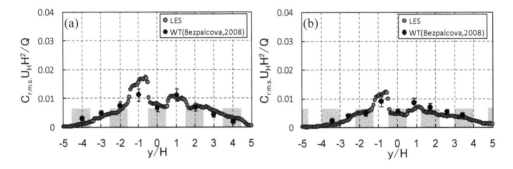

Figure 7. Spanwise profiles of RMS concentration at a height of $0.29H$. **(a)** at the 4th row behind source location in the case of $\lambda_f = 0.25$. **(b)** at the 5th row behind source location in the case of $\lambda_f = 0.33$.

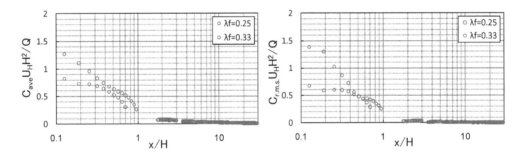

Figure 8. Streamwise variations of mean and RMS concentrations at plume axis at a height of $0.29H$.

high concentration region around $y/H = 0.0$ and the decrease towards the plume edge is the same as for the experimental data. The RMS concentration profile of the experiment shows the local minimum at $y/H = 0.0$ and the local maximum around $y/H = -1.0$ and 1.0. Although LES data over-predict around $y/H = -1.0$ slightly, the shape of the RMS concentration profile of LES is the same as for the experiment. From these results, the mean and RMS concentrations of LES are found to be generally similar in magnitude to that of the experiment. Therefore, it is considered that our LES model for plume dispersion within a building array gives satisfactory results. Both concentration characteristic mean and RMS show an asymmetric pattern for LES data. The maximum value can be found at the left hand side ($y/H < 0$). This

is due to even number of the computational cells and asymmetric placement of the source described in Sect. 3.2.

Figure 8 shows the streamwise variation of mean and RMS concentrations at $y/H = 0.0$ at a height of $0.29H$. At the shorter distances from the point source, $x/H < 1.0$, plume dispersion is enhanced by each building. Therefore, the mean concentration in $\lambda_f = 0.33$ becomes smaller than that in $\lambda_f = 0.25$. At the position located away from the point source, $x/H > 1.0$, the magnitude of the decrease with downwind distance becomes small and these data become quite similar due to the sheltering effect by the building array. Macdonald et al. (1997) investigated the influence of obstacle density on mean concentration of a plume by the field experiments. They mentioned that lateral dispersion is enhanced

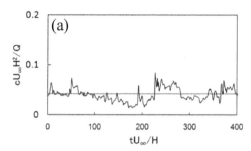
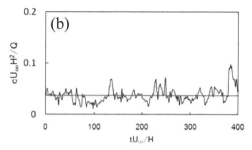

Figure 9. Time series of concentarion fluctuation at the central street canyon. (**a**) $\lambda_f = 0.25$. (**b**) $\lambda_f = 0.33$. Red line indicates average level.

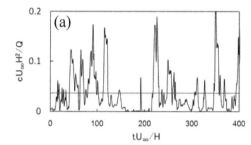
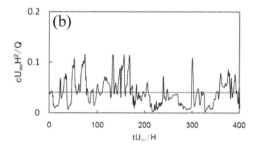

Figure 10. Time series of concentarion fluctuation at the crossing section. (**a**) $\lambda_f = 0.25$. (**b**) $\lambda_f = 0.33$. Red line indicates average level.

in the denser arrays for short distances from a releasing point but is generally similar to the open-terrain case for larger distances due to the sheltering effect by the array. The tendency to decrease with downwind distance depending on obstacle density is similar the field experimental study of Macdonald et al. (1997).

The RMS concentration in $\lambda_f = 0.33$ also becomes smaller than that in $\lambda_f = 0.25$ at $x/H < 1.0$ due to the smoothing of concentration fluctuations by the smaller turbulent eddy with denser arrays. At $x/H > 1.0$, these data are quite similar due to the homogeneous mean concentration field in both cases.

4.3 Characteristics of the peak concentration

In case of accidental or intentional release of toxic or flammable gases into the atmosphere, it is important to estimate not only the mean but also the instantaneous high concentrations. In this section, we first investigate time series of concentration fluctuation and then discuss the characteristics of the peak concentrations. Figures 9 and 10 show time series of concentration fluctuation at the central street canyon and crossing section at the 4th and 5th row behind source location in cases of $\lambda_f = 0.25$ and 0.33, respectively. In case of $\lambda_f = 0.25$, concentrations fluctuate smoothly and continuously at the position of central street canyon, while instantaneous high concentrations which exceed the average level frequently occur at the crossing section. Mavroidis and Griffiths (2001) examined time series of concentration fluctuation under different conditions, such as in open-terrain, behind an isolated cube and within a building array by the

wind tunnel experiments. It was shown that, particularly, higher concentration peaks in a gap between two cubes occur much larger than those behind a cube within a building array. These patterns of concentration fluctuation inside and outside of the cavity region of a building are similar to the wind tunnel experiment by Mavroidis and Griffiths. On the other hand, in case of $\lambda_f = 0.33$, concentrations are found to fluctuate around the average level both at the central street canyon and crossing section.

Figure 11 shows probability distribution functions ($1 - p(c)$) of concentration fluctuation at the central street canyon and crossing section. The probability distribution functions in cases of $\lambda_f = 0.25$ and 0.33 are found to be almost the same at the central street canyon among the cases. On the other hand, at the crossing section, instantaneous high concentrations in case of $\lambda_f = 0.25$ occur much more frequently than those in case of $\lambda_f = 0.33$. Furthermore, we evaluate the peak value c_{99} defined as the values determined from $1 - p(c) = 0.99$ in the wind tunnel experiment and LES. The peak concentration ratios (c_{99}/C_{ave}) of LES at the central street canyon are 2.0 and 2.2 in $\lambda_f = 0.25$ and 0.33, while those of the experiment are 2.0 and 1.9 in $\lambda_f = 0.25$ and 0.33. At the crossing section, the peak ratios of LES are 4.5 and 2.7 in $\lambda_f = 0.25$ and 0.33, while those of the experiment are 2.6 and 2.2 in $\lambda_f = 0.25$ and 0.33. Although LES data at the central street canyon are in good agreement with the experimental data, those at the crossing position are overestimated.

Focusing on the obstacle density effect on the peak ratios, it is found from the experimental data that the peak ratio at the crossing position becomes much larger than that at the

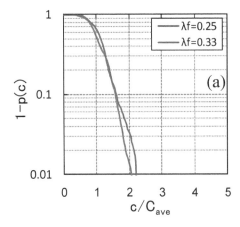 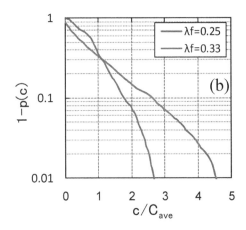

Figure 11. Probability distribution function of concentarion fluctuation for different roughness density. (**a**) at central street canyon. (**b**) at crossing section.

central street canyon in $\lambda_f = 0.25$ while the peak ratios are similar in both locations in $\lambda_f = 0.33$. This tendency is similar to the results of our LES model.

From these results, it is obvious that the peak concentration ratios show highly different values depending on the locations and obstacle density. This fact indicates that, for the assessment of human health hazard or the safety analysis of the hazardous gas within urban areas, it is important to locally evaluate the peak concentrations considering obstacle morphology.

5 Conclusions

In this study, we perform LES of plume dispersion within building arrays with large obstacle densities which corresponds to densely built-up urban areas and investigate mean and fluctuating concentrations. The obtained results are as follows:

1. The approach flow is generated by incorporating existing inflow turbulence generation method into an upstream small part of the driver region with a tripping fence and roughness blocks. By this turbulence generation method, the approach flow corresponding to a neutral atmospheric boundary layer is considered to be obtained.

2. When compared to the experimental results of Bezpalcova and Ohba (2008), the spanwise profiles of mean and RMS concentrations are generally similar in the magnitude to the experimental data. Therefore, it is considered that our LES model for plume dispersion within a building array gives satisfactory results and can be used for deeper and finer investigation of the flow and concentration field within the array.

3. The influence of obstacle density on mean concentration is very large for $x/H < 1.0$ due to the enhancement of plume spreads by each building. For $x/H > 1.0$, its influence becomes small and these data are quite similar due to the sheltering effect by the building array. RMS concentration becomes smaller with denser arrays for $x/H < 1.0$. However, these data are quite similar due to the homogeneous mean concentration field in both cases for $x/H > 1.0$.

4. Although LES peak concentration ratios at the central street canyon are in good agreement with the experimental data, those at the crossing position are overestimated. However, the patterns of the peak ratios depending on the locations and obstacle density are similar to the wind tunnel experiment. Focusing on the obstacle density effect on the peak ratios, it is found that those show different values depending on the locations and obstacle density.

These results imply that, for the assessment of human health hazard or the safety analysis of the hazardous gas within urban areas, it is important to locally evaluate the peak concentrations considering urban morphology.

Although the comparison of the LES data with the experimental data is not sufficiently discussed, we attempted to examine the influence of obstacle density on the spatial distribution of concentrations and the peak concentration characteristics. In order to quantitatively evaluate the obstacle density effects on the dispersion characteristics, the prediction accuracy of our LES model should be further investigated and improved.

Acknowledgements. The study was supported by Japan Society for Promotion of Science, KAKENHI 22710051.

Edited by: M. Piringer
Reviewed by: two anonymous referees

References

Bezpalcova, K. and Ohba, M.: Advective and turbulent vertical fluxes of the passive contaminant inside an urban canopy, Proceeding of 20th National Symposium on Wind Engineering, Tokyo, Japan, 20, 19–24, 2008.

Counihan, J.: Adiabatic atmospheric boundary layers: a review and analysis of data from the period 1880-1972, Atmos. Environ., 9, 871–905, 1975.

Davidson, M. J., Snyder, W. H., Lawson, R. E., and Hunt, J. C. R.: Wind tunnel simulations of plume dispersion through groups of obstacles, Atmos. Environ., 30, 3715–3731, 1996.

Engineering Science Data Unit: Characteristics of atmospheric turbulence near the ground Part2 Single point data for strong winds (neutral atmosphere), ESDU Item 85020, 1985.

Goldstein, D., Handler, R., and Sirovich, L.: Modeling a no-slip flow boundary with an external force field, Journal of Computer Physics, 105, 354–366, 1993.

Gresho, P. M.: Some interesting issues in incompressible fluid dynamics, both in the continuum and in numerical simulation, Adv. Appl. Mech., 28, 45–140, 1992.

Harlow, F. and Welch, J. E.: Numerical calculation of time-dependent viscous incompressible flow of fluid with a free surface, Phys. Fluids, 8, 2182–2189, 1965.

Kataoka, H. and Mizuno, M.: Numerical flow computation around aeroelastic 3D square cylinder using inflow turbulence, Wind Struct., 5, 379–392, 2002.

Macdonald, R., Griffiths, R., and Cheah, S.: Field experiments of dispersion through regular arrays of cubic structures, Atmos. Environ., 31, 6, 783–795, 1997.

Mavroidis, I. and Griffiths, R.: Local characteristics of atmospheric dispersion within building arrays, Atmos. Environ., 35, 2941–2954, 2001.

Murakami, S., Mochida, A., and Hibi, K.: Three-dimensional numerical simulation of air flow around a cubic model by means of large eddy simulation, J. Wind Eng. Ind. Aerod., 25, 291–305, 1987.

Shi, R. F., Cui, G. F., Wang, Z. S., Xu, C. X., and Zhang, Z. S.: Large eddy simulation of wind field and plume dispersion in building array, Atmos. Environ., 42, 1083–1097, 2008.

Smagorinsky, J.: General circulation experiments with the primitive equations, Mon. Weather Rev., 91, 3, 99–164, 1963.

Takewaki, H., Nishiguchi, A., and Yabe, T.: Cubic Interpolated Pseudo-particle method (CIP) for solving hyperbolic-type equations, Journal of Computer Physics, 61, 261–268, 1985.

Van Driest, E. R.: On turbulent flow near a wall, Journal of Aerospace Science, 23, 1007–1011, 1956.

19

Solar irradiance in clear atmosphere: study of parameterisations of change with altitude

A. Oumbe[1], **Ph. Blanc**[2], **B. Gschwind**[2], **M. Lefevre**[2], **Z. Qu**[2], **M. Schroedter-Homscheidt**[1], **and L. Wald**[2]

[1]German Remote Sensing Data Center, German Aerospace Center (DLR), Oberpfaffenhofen, Germany
[2]Centre for Energy and Processes, MINES ParisTech, BP 204, 06905 Sophia Antipolis cedex, France

Abstract. Parameterisation of changes of the solar irradiance at ground level with a specific variable (e.g. solar zenithal angle, aerosol optical depth, altitude, etc.) is often used in operational processes because it saves computational time. This paper deals with the modelling of the vertical profile of downwelling solar irradiance for the first two kilometres above ground in clear sky conditions. Two analytical parameterisations are evaluated for direct and global irradiance in spectral bands as well as for the total irradiance. These parameterisations reproduce the vertical profile with good accuracy for global spectral irradiance and are less accurate for direct component, especially in turbid atmosphere. A piecewise linear interpolation technique using irradiance values known at surface and 4 altitudes every 500 m performs better in any case.

1 Introduction

Satellite-derived assessments of surface downwelling solar irradiance (SSI) are more and more used by engineering companies in solar energy. Performances are judged satisfactory for the time being. Nevertheless, requests for more accuracy are increasing, in particular in the spectral definition and in the decomposition of the global radiation into its direct and diffuse components. One approach to reach this goal is to improve both the modelling of the radiative transfer and the quality of the inputs describing the optical state of the atmosphere. Within their joint project Heliosat-4, DLR and MINES ParisTech have adopted this approach aiming at creating advanced databases of solar irradiance succeeding to the current ones HelioClim and SolEMi (Oumbe et al., 2009). The Heliosat-4 method will be applied to Meteosat images.

This article contributes to this project and its practical implementation. It deals with clear-sky conditions and focuses on the modelling of the change in the SSI with altitude z for the first two kilometres above ground level at z_0. In operations, the SSI is assessed at the mean elevation of the Meteosat pixel. To answer users' requests, we need to provide the SSI at a different elevation whenever the average pixel elevation is different from that of the considered site. This issue is similar to the modelling of the dependency of irradiance with altitude in the cloud-free atmosphere (Abdel Wahab et al., 2009; Gueymard and Thevenard, 2009). The SSI is then set to the irradiance in the free atmosphere for the corresponding elevation z. It would be possible to run a radiative transfer model (RTM) to obtain such a z-profile. However, running a RTM is very time-consuming and an analytical function or a simple linear interpolation technique would be preferable from an operational point of view (Mueller et al., 2009). Therefore, the problem statement is: can we model the vertical profile up to 2 km with a sufficient accuracy using fast parameterisations or linear interpolators?

A few parameterisations have been published that may answer this question (Abdel Wahab et al., 2009; Gueymard and Thevenard, 2009; Oumbe and Wald, 2009). However, they have been studied in a limited number of cases. This article establishes the performances of two parameterisations for all clear-sky conditions and for the spectral distribution of the irradiance by comparing their outcomes to those from a RTM considered as delivering reference values. In addition, we compare them to two standard interpolation techniques in order to assess their potential benefits.

Both parameterisations require as inputs solar irradiances at two different altitudes, here 0 km and 2 km. These irradiances are provided by the RTM libRadtran (www.libradtran.org). Aside its use to obtain these two irradiances, libRadtran is a reference against which we assess the performances of each parameterisation in retrieving irradiance for 32 spectral bands and total. The spectral resolution used here is the

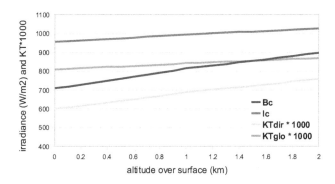

Figure 1. Example of change in irradiance (B_c for direct and I_c for global) and clearness index (KT) with altitude. KTdir and KTglo are direct and global KT.

correlated k-approximation (Kato et al., 1999), for a compromise between accuracy of results and speed of execution. With this method, atmospheric transmittance is computed following the spectral absorption of molecules in the atmosphere. The 32 spectral bands span from 240 nm to 4600 nm. The SSI is null in bands #1 and #2.

2   Parameterisations

Figure 1 illustrates the change in solar downwelling irradiance with altitude z above the surface ranging from 0 km to 2 km. In this example, the irradiances were computed with libRadtran using standard conditions in Europe, a solar zenithal angle equal to 30° and an aerosol optical depth at 550 nm equal to 0.3. The shape of the two curves, B_c for direct and I_c for global, is not the same: the direct irradiance increases faster with altitude than the global irradiance, due to the higher AOD sensitivity of direct irradiance.

The clearness index KT is the ratio of the irradiance at ground level to that at the top of atmosphere. Expectedly, the variation of KT (KTdir, KTglo) with altitude is similar to that of SSI (Fig. 1). Actually, in this work, we always compute SSI, with libRadtran and with the chosen parameterisation, and derive KT from it. Using KT is a mean to show how good the method reproduces the effect of atmospheric contents on solar downwelling spectral irradiance, since the errors in SSI are expressed relatively to the irradiance at the top of atmosphere. Using KT instead of SSI is more convenient in several cases and both should be used. For example, the magnitude of the SSI changes from one Kato band to another and it is difficult to compare the deviation in SSI for different Kato bands. On the other side, errors in KT are commensurate and one can analyse and compare the errors for all cases more efficiently. Moreover, the parameterisations are explicitly products of extraterrestrial irradiance (I_0) and other parameter, corresponding to KT. They are based on the assumption that the vertical profile of irradiance follows

an exponential form. This assumption is supported by the law of Beer-Bouguer-Lambert for the direct component.

2.1   Parameterisation 1

The parameterisation 1 (p1) describes the global and direct irradiances with the following functions (Oumbe and Wald, 2009):

$$I_c(z) = I_0(1 - A(z_0) \exp[-\alpha(z - z_0)])$$
$$B_c(z) = I_0(1 - A_B(z_0) \exp[-\alpha_B(z - z_0)]) \qquad (1)$$

Knowing clear-sky irradiances at two different altitudes, $I_c(z_H)$ at z_H and $I_c(z_0)$ at z_0, $A(z_0)$ and α are determined:

$$A(z_0) = 1 - (I_c(z_0)/I_0)$$
$$\alpha = -\ln[(I_0 - I_c(z_H))/(I_0 - I_c(z_0))]/(z_H - z_0) \qquad (2)$$

The same equations hold for $A_B(z_0)$ and α_B, where $B_c(z_H)$ and $B_c(z_0)$ are the direct clear-sky irradiances at z_H and z_0.

2.2   Parameterisation 2

The second parameterisation (p2) is an extension of that proposed by Abdel Wahab et al. (2009) and inspired from Gueymard and Thevenard (2009). Here, global and direct irradiances are defined as:

$$I_c(z) = I_0 \exp[-\tau(z_0)\beta^{(z0-z)}]$$
$$B_c(z) = I_0 \exp[-\tau_B(z_0)\beta_B^{(z0-z)}] \qquad (3)$$

Knowing clear-sky irradiances at two different altitudes, $I_c(z_H)$ at z_H and $I_c(z_0)$ at z_0, $\tau(z_0)$ and β are given by:

$$\tau(z_0) = -\ln(I_c(z_0)/I_0)$$
$$\beta = \exp[\ln(\tau(z_0)/\tau(z_H))/(z_0 - z_H)] \qquad (4)$$

The same equations hold for $\tau_B(z_0)$ and β_B, where $B_c(z_H)$ and $B_c(z_0)$ are the direct clear-sky irradiances at z_H and z_0.

2.3   Linear interpolations

The interpolator p3 is a simple linear interpolation between the irradiances known at 0 km and 2 km and extrapolation for other altitudes. The interpolator p4 is a piecewise linear interpolation. Four intervals are defined, from 0 km to 2 km, every 500 m. The irradiance is known at each of the five limits. Within each interval, the irradiance is computed by a linear interpolation; extrapolation is performed for other altitudes.

3   Comparison with libRadtran

A Monte-Carlo technique is applied to randomly select 5000 sets within the 9D-space defined by discrete values taken by the 9 most prominent inputs to libRadtran with respect SSI: solar zenith angle, aerosol optical depth, type and Angstrom

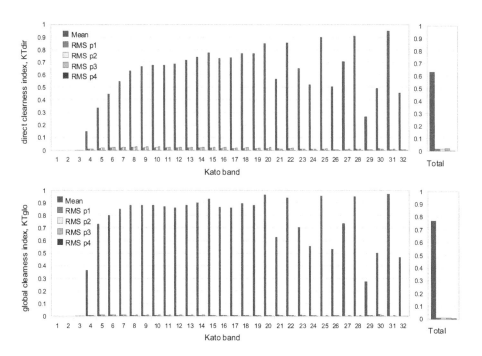

Figure 2. Mean KT and RMS for each parameterisation.

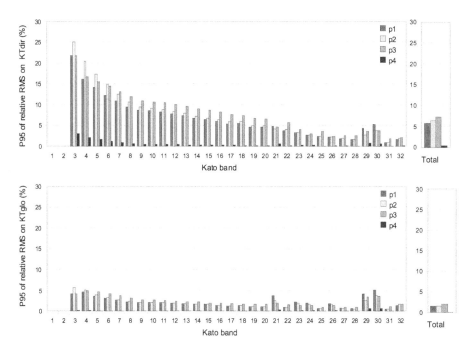

Figure 3. P95 of relative RMS for each parameterisation.

exponent, total column water, total column ozone, altitude of the ground, ground albedo and atmospheric profile. We have selected these optical properties considering their observed marginal distribution. More precisely, we have chosen the uniform distribution as a model for marginal probability for all parameters except for the parameters related to aerosol, albedo and ozone. We have selected the normal law for the Angstrom coefficient, the gamma law for the aerosol optical depth and the beta law for total column ozone and ground albedo. The parameters of the laws are empirically determined from the analyses of the observations made in the AERONET network for aerosol properties and from meteorological satellite-based ozone products. The aerosol type "continental average" is used half the time. By weighting

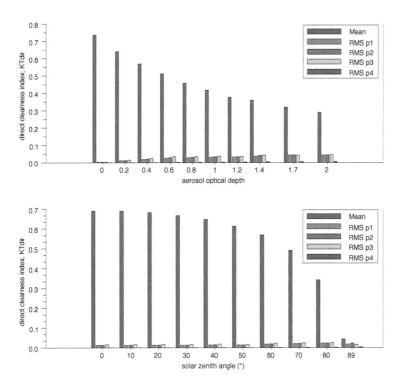

Figure 4. Variation of KTdir and of corresponding RMS with aerosol optical depth and solar zenith angle. Spectral and total KT are mixed up.

these selections, we reduce the amount of rare atmospheric states where very few SSI and large relative errors are obtained. For each 9-tuple, libRadtran outputs are total global and direct irradiances, and their spectral values for the 32 Kato spectral bands, for 21 altitudes from 0 to 2 km every 100 m above the surface.

The parameterisations p1, p2 and p3 are adjusted on the irradiances at 0 km and 2 km. Three more altitudes are used with p4. Irradiances at altitudes not used are computed with each parameterisation p1 to p4 and compared to those from libRadtran. The comparison is made on the clearness index KT defined as the ratio between the spectral surface irradiance and the corresponding spectral top-of-atmosphere. KTdir and KTglo are computed. The differences are synthesized by the RMS (Root Mean Square deviation) and the percentile 95 % (P95) of the frequency distribution of the RMS within the 21 altitudes. They are compared to the mean KT (Fig. 2).

Errors in KTglo are very small for each band and total: each parameterisation reproduces well the vertical profile. As for the total, the errors in KTdir are greater in absolute value than those in KTglo for each band. Their relative values are also greater: they may amount to 3–5 % at short wavelength and 1–2 % at long wavelength. The parameterisation p3 performs the worst of the four tested functions. The parameterisation p1 and p2 are close, their accuracies are similar on KTglo, p1 performs better than p2 at wavelengths shorter than 800 nm, and p2 performs better at longer wavelengths. The parameterisation p4 performs the best.

Figure 3 shows the P95 and stresses the inaccuracies that can occur in very turbid atmospheres.

P95 for KTglo is less than 5 % for all parameterisations and for most wavelengths. P95 is less than 10 % for KTdir for all parameterisations for wavelengths greater than 600 nm, and reaches a maximum of 25 % at 300 nm for p2. This demonstrates that these parameterisations are accurate for the global and direct irradiances. Note that P95 in KTdir and KTglo for p4 is very low for all bands. Relative performances of parameterisations are the same as in Fig. 2. p3 remains the worst and p4 the best. p1 is better than p2 at short wavelengths and p1 is better at longer wavelengths. If one selects the best parameterisation between p1 and p2 at each Kato band, P95 will be less than 5 % for KTglo for all wavelengths and for KTdir for all wavelengths greater than 800 nm. P95 greater than 10 % will be obtained only for wavelengths less than 450 nm. These regions are not the main contributors to the total extraterrestrial irradiance: they represent only 15 %, and the conclusion should take this into account of total irradiance. Accuracy of p4 remains good; the error is the greatest in the less-energetic bands and is less than 10 % for all bands. Actually, these large relative errors in KT are small in SSI: the corresponding RMS is less than 6 W m^{-2} and 8 W m^{-2} on direct and global total irradiance.

The errors for each parameterisation are not related to precipitable water and ground albedo but depend on solar zenith angle and aerosol optical depth (Fig. 4). The RMS increases with the solar zenith angle or aerosol optical depth, and therefore as KTdir decreases. This does not lead to an important absolute RMS, but its relative value can be greater than 50 %, especially when both solar zenith angle and aerosol optical depth are large. In such extreme cases, only the parameterisation p4 performs well. The RMS increases also with Angstrom alpha coefficient, especially for large wavelengths. Similar observations are made for the bias.

4 Conclusions

This study demonstrates that it is possible to reproduce the vertical profile of the spectral global and direct irradiances under clear-sky with sufficient accuracy using an analytical parameterisation. Such parameterisations can be used in the case of extrapolating irradiance assessed for a pixel to a site of different altitude within this pixel, or extrapolating irradiance measured at a station to a close site of different altitude, or for designing a fast method for processing satellite data based on RTM.

In low atmospheric turbidity, all four assessed parameterisations are accurate. The accuracy decreases as the radiation path length in the atmosphere (turbidity and solar zenith angle) increases, especially for direct irradiance. For standard atmospheric compositions, p1 performs well for wavelengths shorter than 800 nm and p2 performs well for longer wavelengths. They could be good candidates for the regions in the field-of-view of Meteosat where the aerosol optical depth is rarely greater than 1, with a mean close to 0.3–0.4. Other cases are rare and correspond to very small irradiances. In such extreme cases, high relative deviations are obtained with p1, p2 and p3. If one looks for a robust modelling, one should select the piecewise linear interpolation.

Acknowledgements. The research leading to these results has received funding from the European Union's Seventh Framework Programme (FP7/2007-2013) under Grant Agreement no. 218793 (MACC project). The authors would also like to thank the developers of libRadtran (http://www.libradtran.org).

Edited by: S.-E. Gryning
Reviewed by: two anonymous referees

References

Abdel Wahab, M., El-Metwally, M., Hassan, R., Lefèvre, M., Oumbe, A., and Wald, L.: Assessing surface solar irradiance in Northern Africa desert climate and its long-term variations from Meteosat images, Int. J. Remote Sens., 31, 261–280, 2009.

Gueymard, C. A. and Thevenard, D.: Monthly average clear-sky broadband irradiance database for worldwide solar heat gain and building cooling load calculations, Sol. Energy, 83, 1998–2018, 2009.

Kato, S., Ackerman, T., Mather, J., and Clothiaux, E.: The k-distribution method and correlated-k approximation for shortwave radiative transfer model, J. Quant. Spectrosc. Ra., 62, 109–121, 1999.

Mueller, R. W., Matsoukas, C., Gratzki, A., Behr, H. D., and Hollmann, R.: The CM-SAF operational scheme for the satellite based retrieval of solar surface irradiance – A LUT based eigenvector hybrid approach, Remote Sens. Environ., 113, 1012–1024, 2009.

Oumbe, A. and Wald, L.: A parameterisation of vertical profile of solar irradiance for correcting solar fluxes for changes in terrain elevation, in: Earth Observation and Water Cycle Science Towards a Water Cycle Multi-mission Observation Strategy, 18–20 November at ESRIN, Frascati, Italy, 2009.

Oumbe, A., Blanc, Ph., Ranchin, T., Schroedter-Homscheidt, M., and Wald, L.: A new method for estimating solar energy resource, in: Proceedings of the ISRSE 33, Stresa, Italy, 4–9 May 2009, published by Joint Research Center, Ispra, Italy, USBKey, paper 773, 2009.

20

Correlation between air-sea heat fluxes over the Aegean Sea and the total precipitable water over Europe and North Africa

V. P. Papadopoulos[1], A. Bartzokas[2], T. Chronis[1], S. Ruiz[3], N. Zarokanellos[1], and G. Ferentinos[4]

[1]Hellenic Centre for Marine Research, Anavissos, Greece
[2]Department of Physics, University of Ioannina, Greece
[3]IMEDEA (CSIC-UIB), Mallorca, Spain
[4]Department of Geology, University of Patras, Greece

Abstract. The relation between the air-sea heat fluxes in the Aegean Sea and the total precipitable water (TPW) over Europe and North Africa is investigated. An one-point linear correlation between the components of surface heat exchange and TPW is employed. During winter and for the shortwave radiation, a dipole of opposite correlation (see-saw teleconnection), is observed between the northeast Europe and the East Mediterranean Sea. This pattern is inverted for the longwave radiation and especially during the summer is limited to a strong positive south pole. Both spatial correlation patterns underline the influence of cloudiness during the winter season and the specific humidity during summer on the radiative fluxes behavior. Regarding the turbulent air-sea heat flux components – latent and sensible heat – the correlation pattern is enhanced. The winter pattern exhibits positive correlation over Europe with the highest values ($r > 0.75$) over the Balkan Peninsula. The identified correlation patterns, although they remain dominant, weaken during the summer. The herein findings suggest that the stronger relation between the TPW and the turbulent fluxes is regulated by the wind regime.

1 Introduction

Air-sea heat fluxes are critical for the climate regulation and affect a number of significant processes for both ocean and atmosphere from the global to the local scale. They govern the oceanic and atmospheric circulation and they are important inputs in the numerical weather prediction (NWP) models.

The total precipitable water (TPW) is an atmospheric parameter related to the ocean-atmosphere energy balance. The TPW is highly correlated with cloud formation and precipitation and is mostly used in the precipitation forecasting/nowcasting. Moreover, it exhibits a positive feedback to the global warming as higher water vapor amount in the atmospheric column increases the fraction of the infrared radiation which is trapped within the atmosphere (IPCC, 2001).

The Aegean Sea is an important area where intense air-sea interactions lead to the formation of intermediate and deep waters. The latter affect the thermohaline circulation of the Eastern Mediterranean Sea (Roether et al., 1996; Nittis et al., 2003). This vital process for the renewal and oxygenation of near bottom water layers is directly related to the surface heat exchange regime. The study at hand examines the relationship between the air-sea heat fluxes over the Aegean Sea and the spatial distribution of the TPW over the Northeast Atlantic, Europe and North Africa. The two parameters are regulated by common factors like the sea level pressure (SLP) variability and the wind field. Furthermore, cloudiness and specific humidity, both strongly linked to TPW, modulate the radiative components (shortwave and longwave radiation, SW and LW). Conversely, evaporation releases latent heat (LH) and transfers water from ocean to atmosphere and the turbulent fluxes affect the precipitation regime (Lolis et al., 2004). Therefore, the co-variability of the TPW and the surface fluxes is expectedly of high interest for the climatology of the Aegean Sea and the Eastern Mediterranean.

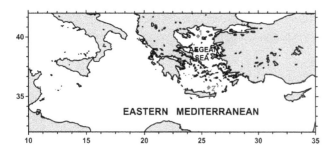

Figure 1. Location of the two sites in the North (**1**) and South (**2**) Aegean Sea at which the surface heat exchanges are correlated with the field of TPW.

2   Data and methodology

Monthly mean air-sea heat fluxes at two sites in the North and the South Aegean Sea (Fig. 1) originate from two different datasets. The special Mediterranean HIPOCAS dataset (1958–2001, Sotillo et al., 2005; Ratsimandresy et al., 2008), provided by Puertos del Estado (Spain), is employed for the radiative, whereas the global OAFlux (1958–2006, Yu and Weller, 2007) for the turbulent fluxes (latent and sensible heat, LH and SH). The two datasets are qualified as statistically robust for the Aegean Sea region after a comparison with an observational dataset (Papadopoulos et al., 2010). Note that we use the oceanographic convention for the heat flux according which positive values represent heat gain by the sea.

Monthly mean TPW values are retrieved from the ERA-40 reanalysis dataset (Uppala et al., 2005) and cover an area extending from 15 W to 45 E and from 25 N to 70 N at a 2.5 degrees resolution square grid. In order to examine the relation between the surface heat exchanges and the TPW field, we apply one-point linear correlation between the air-sea heat flux anomalies and the TPW anomalies at each of the 21 by 19 total grid points. This correlation is computed for both Aegean Sea sites and separately for the cold (November–March) and the warm (May–September) season of the year. A composite correlation map for the North and South Aegean is presented for each period of the year and for each flux component of the air-sea heat flux. Every composite correlation map is produced by averaging the correlation maps for the North and South Aegean Sea. Only the statistically significant (r) values at a confidence level of 95% are accounted for. Especially for the SH flux, we present only the winter correlation. Usually, during summer SH displays positive values but often negative monthly mean values are also observed. The sign of SH is regulated by the difference between the air and the sea temperature, whereas the wind speed determines its magnitude. For this reason, including positive and negative SH values misrepresents the evaluation of the calculated correlation coefficients.

3   Results and discussion

Figure 2 illustrates the correlation maps for the cold and the warm season of the year and for the radiative and turbulent (LH only) components. During winter, SW and TPW exhibit a dipole of opposite correlation (see-saw teleconnection, Fig. 2a), between the north Europe and East Mediterranean Sea. Since low SW values go along with high LW, this pattern is inverted for the LW (Fig. 2c). SW and LW are inversely influenced by the cloud fraction which in turn is strongly correlated with TPW. Overcast conditions with high TPW values implies lower SW values and higher LW. Both winter correlation patterns underline that an increase of TPW over northern Europe and the simultaneous decrease over eastern Mediterranean imply a decrease of total cloud coverage over the Aegean Sea and vice-versa. Composite SLP maps during the lowest and highest SW values also support the observed see-saw pattern. The lowest values of SW over the South Aegean are associated with low SLP over central Mediterranean and high SLP over central and eastern Europe. This SLP pattern favors higher values of TPW over the Aegean Sea and lower over central and eastern Europe (Fig. 3a). On the contrary, during the highest winter SW values high pressures prevail over the greater part of Europe and over the Balkan Peninsula (Fig. 3b). Accordingly to this pattern, west winds cause higher TPW values over north Europe, whereas dry northerlies reduce the cloud fraction and increase SW values over the Aegean Sea. During summer only the negative south pole is observed for the SW (Fig. 2b) and a stronger (in comparison to the winter) positive south pole for the LW (Fig. 2d). The summer cloud fraction is systematically low and the higher specific humidity takes the role of the cloudiness creating a physical trap for the LW emitted by the sea surface. Both, cloud fraction and specific humidity, reduce the heat loss by the LW (see the relevant bulk formula, e.g. Bignami et al., 1995). As higher temperatures favor higher specific humidity values this provides an adequate explanation of the TPW contribution to the positive feedback of the greenhouse warming.

Regarding the turbulent air-sea heat flux components, a different correlation pattern is shown (Fig. 2e and f). These exhibit a strong positive correlation ($r > 0.75$) over the Balkan Peninsula indicating TPW decrease (increase) when turbulent fluxes in Aegean Sea increase (decrease). The flux anomalies decrease when the heat losses increase as LH and SH are both negative during winter. This pattern is also consistent with the SLP composites during the lowest and highest LH values (Fig. 3c and d). High SLP over central Europe favors the strong cold and dry northerlies dominating over the Aegean Sea throughout the year (Chronis et al., 2010) and increases the turbulent heat loss. At the same time, the lower TPW over the continental Europe is attributed to the prevailing high barometric pressures. These high pressures covering the greater part of continental Europe create the necessary pressure gradient for the dominant northerlies over the

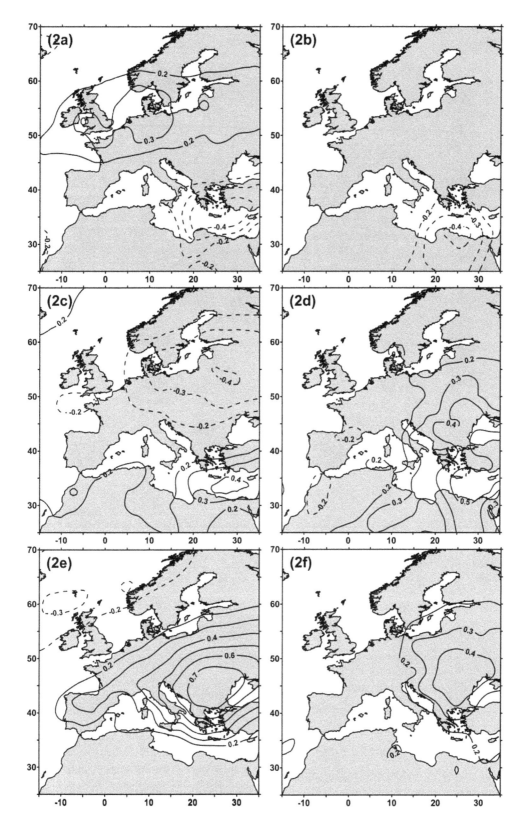

Figure 2. Correlation maps between TPW and (**a**) shortwave radiation during winter, (**b**) shortwave radiation during summer, (**c**) longwave radiation during winter, (**d**) longwave radiation during summer, (**e**) latent heat during winter and (**f**) latent heat during summer. Dashed lines indicate negative correlation.

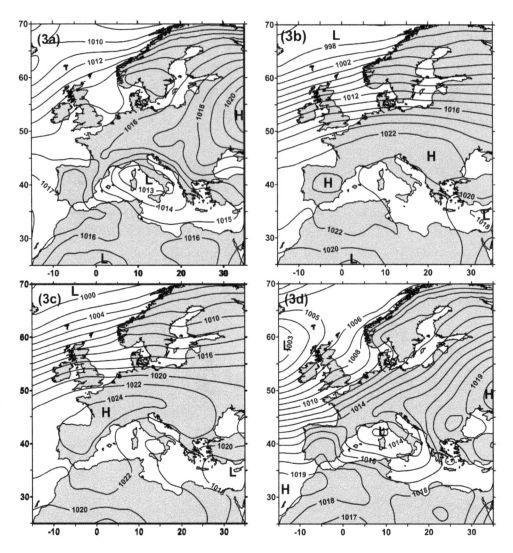

Figure 3. Composite SLP (in hPa) distribution for: **(a)** the lowermost 10% extreme values of SW occurring within the period 1958–2001 in South Aegean during winter, **(b)** the uppermost 10% SW values, **(c)** the lowermost 10% LH values and **(d)** the uppermost 10% LH values.

Aegean. The presence of the Cyprus Low during the winter and the thermal Low of Northwest Asia during the summer is also synergistic in enhancing the northerlies (Flocas et al., 2010). The identified correlation pattern for the latent heat weakens during summer although it still remains prominent (Fig. 2f).

4 Conclusions

The correlation between the components of the air-sea heat fluxes over the Aegean Sea and the TPW over Europe and North Africa reveals the following key features:

– A moderate correlation is observed between the radiative fluxes and the TPW which are naturally related to cloudiness during winter and specific humidity during summer.

– The longwave radiation is the only flux component that exhibits stronger correlation with TPW during summer than winter.

– Higher specific humidity during summer traps a higher fraction of the longwave radiation indicating the role of TPW in positive feedback of the global warming.

– The correlation between turbulent fluxes and TPW ($r > 0.75$) is stronger than the correlation between radiative fluxes and TPW since turbulent fluxes and TPW are regulated by the same parameters.

– The correlation pattern for the turbulent fluxes is affected by the presence of high pressures centers over the continental Europe which in combination with the Cyprus Low during winter and the thermal NW Asia Low during summer generate the cold and dry northerlies over the Aegean Sea.

In summary, the air-sea heat fluxes in the Aegean Sea and the field of TPW over Europe and North Africa are mainly affected by the wind regime which in turn is regulated by the large scale atmospheric circulation. The spatial and temporal variability of both variables is very important not only for the marine environment but for the whole climatic regime of the Eastern Mediterranean region.

Edited by: B. Lalic
Reviewed by: S. Gualdi and two other anonymous referees

References

Bignami, F., Marullo, S., Santoleri, R., and Schiano M. E.: Long-wave radiation budget in the Mediterranean Sea, J. Geophys. Res., 100(C2), 2501–2514, 1995.

Chronis, T., Papadopoulos, V., and Nikolopoulos, E.: QuickSCA-Tob servations of extreme wind events over the Mediterranean and Black Seas during 2000–2008, Int. J. Climatol., doi:10.1002/joc.2213, 2010.

Flocas, H. A., Simmonds, I., Kouroutzoglou, J., Keay, K., Hatzaki, M., Bricolas, V., and Asimakopoulos, D.: On cyclonic tracks over the Eastern Mediterrranean, J. Climate, 23, 5243–5257, 2010.

Lolis, C. J., Bartzokas, A., and Katsoulis, B. D.: Relation between sensible and latent heat fluxes in the Mediterranean and precipitation in the Greek area during winter, Int. J. Climatol., 24, 1803–1816, 2004.

IPCC: Climate Change 2001: The Scientific Basis, Cambridge University Press, 881 pp., 2001.

Nittis, K., Lascaratos, A., and Theocharis, A.: Dense water formation in the Aegean Sea: numerical simulations during the Eastern Mediterranean Transient, J. Geophys. Res., 108(C9), 8120, doi:10.1029/2002JC001352, 2003.

Papadopoulos, V. P., Chronis, T. G., Ruiz, S., Tragou, E., Josey, S., and Ferentinos, G.: Air-sea heat fluxes estimation in the Aegean Sea (Eastern Mediterranean), Rapp. Comm. Int. Mer Medit., 39, p. 156, 2010.

Ratsimandresy, A. W., Sotilo, M. G., Carretero, J. C., Alvarez, E., and Hajji, H.: A 44-year-high-resoloution ocean and atmospheric hindcast for the Mediterraneaan Basin developed within the HIPOCAS Project, Coast. Eng., 55, 827–842, 2008.

Roether, W., Manca, B., Klein, B., Bregant, B., Georgopoulos, D., Beitzel, V., Kovacevic, V., and Luchetta, A.: Recent changes in eastern Mediterranean deep waters, Science, 271, 333–335, 1996.

Sotillo, M. G., Ratsimandresy, A. W., Carretero, J. C., Bentamy, A., Valero, F., and Gonzaìlez-Rouco, F.: A high-resolution 44-year atmospheric hindcast for the Mediterranean Basin: contribution to the regional improvement of global reanalysis, Clim. Dyn., 25, 219–236, 2005.

Uppala, M. and 45 co-authors: The ERA-40 re-analysis, Q. J. Roy. Meteor. Soc., 131, 612, 2961–3012, 2005.

Yu, L. and Weller, R. A.: Objectively analyzed air-sea heat fluxes for the global ice-free oceans (1981–2005), B. Am. Meteorol. Soc., 88, 4, 527–539, 2007.

Permissions

All chapters in this book were first published in ASR, by Copernicus Publications; hereby published with permission under the Creative Commons Attribution License or equivalent. Every chapter published in this book has been scrutinized by our experts. Their significance has been extensively debated. The topics covered herein carry significant findings which will fuel the growth of the discipline. They may even be implemented as practical applications or may be referred to as a beginning point for another development.

The contributors of this book come from diverse backgrounds, making this book a truly international effort. This book will bring forth new frontiers with its revolutionizing research information and detailed analysis of the nascent developments around the world.

We would like to thank all the contributing authors for lending their expertise to make the book truly unique. They have played a crucial role in the development of this book. Without their invaluable contributions this book wouldn't have been possible. They have made vital efforts to compile up to date information on the varied aspects of this subject to make this book a valuable addition to the collection of many professionals and students.

This book was conceptualized with the vision of imparting up-to-date information and advanced data in this field. To ensure the same, a matchless editorial board was set up. Every individual on the board went through rigorous rounds of assessment to prove their worth. After which they invested a large part of their time researching and compiling the most relevant data for our readers.

The editorial board has been involved in producing this book since its inception. They have spent rigorous hours researching and exploring the diverse topics which have resulted in the successful publishing of this book. They have passed on their knowledge of decades through this book. To expedite this challenging task, the publisher supported the team at every step. A small team of assistant editors was also appointed to further simplify the editing procedure and attain best results for the readers.

Apart from the editorial board, the designing team has also invested a significant amount of their time in understanding the subject and creating the most relevant covers. They scrutinized every image to scout for the most suitable representation of the subject and create an appropriate cover for the book.

The publishing team has been an ardent support to the editorial, designing and production team. Their endless efforts to recruit the best for this project, has resulted in the accomplishment of this book. They are a veteran in the field of academics and their pool of knowledge is as vast as their experience in printing. Their expertise and guidance has proved useful at every step. Their uncompromising quality standards have made this book an exceptional effort. Their encouragement from time to time has been an inspiration for everyone.

The publisher and the editorial board hope that this book will prove to be a valuable piece of knowledge for researchers, students, practitioners and scholars across the globe.

List of Contributors

S. Krähenmann
Institute for Atmospheric and Environmental Sciences, Goethe-University Frankfurt, Frankfurt am Main, Germany

B. Ahrens
Institute for Atmospheric and Environmental Sciences, Goethe-University Frankfurt, Frankfurt am Main, Germany

B. Chimani
Central Institute for Meteorology and Geodynamics, Vienna, Austria

R. Böhm
Central Institute for Meteorology and Geodynamics, Vienna, Austria

C. Matulla
Central Institute for Meteorology and Geodynamics, Vienna, Austria

M. Ganekind
Central Institute for Meteorology and Geodynamics, Vienna, Austria

L. Foresti
Institute of Geomatics and Analysis of Risk, University of Lausanne, Switzerland

M. Kanevski
Institute of Geomatics and Analysis of Risk, University of Lausanne, Switzerland

A. Pozdnoukhov
National Centre for Geocomputation, National University of Ireland Maynooth, Ireland

I. Esau
G.C. Rieber Climate Institute of the Nansen Environmental and Remote Sensing Center, Thormohlensgt. 47, 5006, Bergen, Norway
Bjerknes Centre for Climate Research, Bergen, Norway

S. Zilitinkevich
G.C. Rieber Climate Institute of the Nansen Environmental and Remote Sensing Center, Thormohlensgt. 47, 5006, Bergen, Norway
Division of Meteorological Research, Finnish Meteorological Institute, Helsinki, Finland
Division of Atmospheric Sciences and Geophysics, Department of Physics, University of Helsinki, Finland

C. Simolo
Institute of Atmospheric Sciences and Climate, National Research Council (ISAC-CNR), Bologna, Italy

M. Brunetti
Institute of Atmospheric Sciences and Climate, National Research Council (ISAC-CNR), Bologna, Italy

M. Maugeri
Dipartimento di Fisica, Universit`a degli Studi di Milano, Italy
Institute of Atmospheric Sciences and Climate, National Research Council (ISAC-CNR), Bologna, Italy

T. Nanni
Institute of Atmospheric Sciences and Climate, National Research Council (ISAC-CNR), Bologna, Italy

E. N. Voskresenskaya and
Marine Hydrophysical Institute, Sevastopol, Ukraine

V. N. Maslova
Marine Hydrophysical Institute, Sevastopol, Ukraine

S. Alijani
Department of Urban Designing, Islamic Azad University, Tehran, Iran

B. Alijani
Center of Excellence for Spatial Analysis of Environmental Hazards, Tarbiat Moallem University, Tehran, Iran

P. Domonkos
Centre for Climate Change (C3), Geography Dept., University Rovira i Virgili, Campus Terres de l'Ebre, C. Bet´ania 5, Tortosa, 43500, Spain

J. Spinoni
Department of Physics, University of Milan, Milan, Italy

M. Brunetti
Department of Physics, University of Milan, Milan, Italy
Institute of Atmospheric Sciences and Climate, National Research Council (ISAC-CNR), Bologna, Italy
now at: JRC-IES, Ispra, Italy

M. Maugeri
Department of Physics, University of Milan, Milan, Italy
Institute of Atmospheric Sciences and Climate, National Research Council (ISAC-CNR), Bologna, Italy
now at: JRC-IES, Ispra, Italy

C. Simolo
Institute of Atmospheric Sciences and Climate, National
Research Council (ISAC-CNR), Bologna, Italy
now at: JRC-IES, Ispra, Italy

M. Y. Luna
Agencia Estatal de Meteorolog´ıa, Leonardo Prieto Castro,
8, 28040 Madrid, Spain

J. A. Guijarro
Agencia Estatal de Meteorolog´ıa, Leonardo Prieto Castro,
8, 28040 Madrid, Spain

J. A. López
Agencia Estatal de Meteorolog´ıa, Leonardo Prieto Castro,
8, 28040 Madrid, Spain

F. Pasi
CNR-IBIMET, Firenze, Italy
Consorzio LaMMa, Sesto F. no, Italy

A. Orlandi
CNR-IBIMET, Firenze, Italy
Consorzio LaMMa, Sesto F. no, Italy

L. F. Onorato
Meteo Hydrological Center, Genova, Italy

S. Gallino
Meteo Hydrological Center, Genova, Italy

M. C. Beltrano
Unit`a di Ricerca per la Climatologia e la Meteorologia
applicate all'Agricoltura, Consiglio per la Ricerca e la
Sperimentazione in Agricoltura, Roma, Italy

S. Esposito
Unit`a di Ricerca per la Climatologia e la Meteorologia
applicate all'Agricoltura, Consiglio per la Ricerca e la
Sperimentazione in Agricoltura, Roma, Italy

L. Iafrate
Centro di Ricerca per lo Studio delle Relazioni tra Pianta
e Suolo, Consiglio per la Ricerca e la Sperimentazione in
Agricoltura, Roma, Italy

J. Wibig
Department of Meteorology and Climatology, University
of Lodz, Poland

A. Perrels
Finnish Meteorological Institute FMI, Erik Palm´enin
Aukio 1, P.O. Box 503, 00101 Helsinki, Finland

Th. Frei
formerly at: MeteoSwiss, Kr¨ahb¨uhlstrasse 58, P.O. Box
514, 8044 Z¨urich, Switzerland

F. Espejo
Spanish State Meteorological Agency AEMET, C/
Leonardo Prieto Castro 8, 28040 Madrid, Spain

L. Jamin
Deutscher Wetterdienst DWD, Frankfurter Str. 135, 63067
Offenbach, Germany

A. Thomalla
Deutscher Wetterdienst DWD, Frankfurter Str. 135, 63067
Offenbach, Germany

Z. Qu
Centre for Energy and Processes, MINES ParisTech,
BP207, 06904, Sophia Antipolis, France

P. Blanc
Centre for Energy and Processes, MINES ParisTech,
BP207, 06904, Sophia Antipolis, France

M. Lefèvre
Centre for Energy and Processes, MINES ParisTech,
BP207, 06904, Sophia Antipolis, France

L. Wald
Centre for Energy and Processes, MINES ParisTech,
BP207, 06904, Sophia Antipolis, France

A. Oumbe
German Remote Sensing Data Center, German Aerospace
Center (DLR), Postfach 1116, 82234 Wessling, Germany

O. Y. Skrynyk
Ukrainian Research Hydrometeorological Institute, Kyiv,
Ukraine

R. I. Chernysh
Ukrainian Research Hydrometeorological Institute, Kyiv,
Ukraine

Y. Y. Hrytsyuk
Ternopil National Pedagogical University, Ternopil,
Ukraine

S. Zhong
Department of Geography, Michigan State University,
East Lansing, Michigan USA

X. Xu
Department of Geography, Michigan State University,
East Lansing, Michigan USA

X. Bian
USDA Forest Service Northern Research Station, East
Lansing, Michigan, USA

W. Lu
Department of Geography, Michigan State University,
East Lansing, Michigan USA

H. Nakayama
Japan Atomic Energy Agency, Ibaraki, Japan

K. Jurcakova
Institute of Thermomechanics, Academy of Sciences of the Czech Republic, Prague, Czech Republic

H. Nagai
Japan Atomic Energy Agency, Ibaraki, Japan

A. Oumbe
German Remote Sensing Data Center, German Aerospace Center (DLR), Oberpfaffenhofen, Germany

Ph. Blanc
Centre for Energy and Processes, MINES ParisTech, BP 204, 06905 Sophia Antipolis cedex, France

B. Gschwind
Centre for Energy and Processes, MINES ParisTech, BP 204, 06905 Sophia Antipolis cedex, France

M. Lefevre
Centre for Energy and Processes, MINES ParisTech, BP 204, 06905 Sophia Antipolis cedex, France

Z. Qu
Centre for Energy and Processes, MINES ParisTech, BP 204, 06905 Sophia Antipolis cedex, France

M. Schroedter-Homscheidt
German Remote Sensing Data Center, German Aerospace Center (DLR), Oberpfaffenhofen, Germany

L. Wald
Centre for Energy and Processes, MINES ParisTech, BP 204, 06905 Sophia Antipolis cedex, France

V. P. Papadopoulos
Hellenic Centre for Marine Research, Anavissos, Greece

A. Bartzokas
Department of Physics, University of Ioannina, Greece

T. Chronis
Hellenic Centre for Marine Research, Anavissos, Greece

S. Ruiz
IMEDEA (CSIC-UIB), Mallorca, Spain

N. Zarokanellos
Hellenic Centre for Marine Research, Anavissos, Greece

G. Ferentinos
Department of Geology, University of Patras, Greece

Printed in the USA
CPSIA information can be obtained
at www.ICGtesting.com
JSHW051447221024
72173JS00006B/1605